Springer Series in Statistics

Springer Series in Statistics

D. F. Andrews and A. M. Herzberg, Data: A Collection of Problems from Many Fields for the Student and Research Worker. xx, 442 pages, 1985.

F. J. Anscombe, Computing in Statistical Science through APL. xvi, 426 pages, 1981.

J. O. Berger, Statistical Decision Theory and Bayesian Analysis, 2nd edition. xiv, 425 pages, 1985.

P. Brémaud, Point Processes and Queues: Martingale Dynamics. xviii, 354 pages, 1981.

K. Dzhaparidze, Parameter Estimation and Hypothesis Testing in Spectral Analysis of Stationary Time Series. xii, 300 pages, 1985.

R. H. Farrell, Multivariate Calculation. xvi, 367 pages, 1985.

L. A. Goodman and W. H. Kruskal, Measures of Association for Cross Classifications. x, 146 pages, 1979.

J. A. Hartigan, Bayes Theory. xii, 145 pages, 1983.

H. Heyer, Theory of Statistical Experiments. x, 289 pages, 1982.

M. Kres, Statistical Tables for Multivariate Analysis. xxii, 504 pages, 1983.

M. R. Leadbetter, G. Lindgren and H. Rootzén, Extremes and Related Properties of Random Sequences and Processes. xii, 336 pages, 1983.

E. B. Manoukian, Modern Concepts and Theorems of Mathematical Statistics. xiv, 156 pages, 1986.

R. G. Miller, Jr., Simultaneous Statistical Inference, 2nd edition. xvi, 299 pages, 1981.

F. Mosteller and D. S. Wallace, Applied Bayesian and Classical Inference: The Case of *The Federalist* Papers. xxxv, 301 pages, 1984.

D. Pollard, Convergence of Stochastic Processes. xiv, 215 pages, 1984.

J. W. Pratt and J. D. Gibbons, Concepts of Nonparametric Theory. xvi, 462 pages, 1981.

L. Sachs, Applied Statistics: A Handbook of Techniques, 2nd edition. xxviii, 706 pages, 1984.

E. Seneta, Non-Negative Matrices and Markov Chains. xv, 279 pages, 1981.

D. Siegmund, Sequential Analysis: Tests and Confidence Intervals. xii, 272 pages, 1985.

V. Vapnik, Estimation of Dependences Based on Empirical Data. xvi, 399 pages, 1982.

K. M. Wolter, Introduction to Variance Estimation. xii, 428 pages, 1985.

Edward B. Manoukian

Modern Concepts and Theorems of Mathematical Statistics

Springer-Verlag
New York Berlin Heidelberg Tokyo

Edward B. Manoukian
Department of National Defence
Royal Military College of Canada
Kingston, Ontario K7L 2W3
Canada

AMS Classification: 62-01

Library of Congress Cataloging in Publication Data
Manoukian, Edward B.
Modern concepts and theorems of mathematical statistics.
(Springer series in statistics)
Bibliography: p.
Includes index.
1. Mathematical statistics. I. Title. II. Series.
QA276.M333 1985 519.5 85-14686

Typeset by Asco Trade Typesetting Ltd., Hong Kong.
Printed and bound by R. R. Donnelley & Sons, Harrisonburg, Virginia.
Printed in the United States of America.

9 8 7 6 5 4 3 2 1

ISBN 0-387-96186-0 Springer-Verlag New York Berlin Heidelberg Tokyo
ISBN 3-540-96186-0 Springer-Verlag Berlin Heidelberg New York Tokyo

This book is dedicated to Tanya and Jacqueline

Preface

With the rapid progress and development of mathematical statistical methods, it is becoming more and more important for the student, the instructor, and the researcher in this field to have at their disposal a quick, comprehensive, and compact reference source on a very wide range of the field of modern mathematical statistics. This book is an attempt to fulfill this need and is encyclopedic in nature. It is a useful reference for almost every learner involved with mathematical statistics at any level, and may supplement any textbook on the subject. As the primary audience of this book, we have in mind the beginning busy graduate student who finds it difficult to master basic modern concepts by an examination of a limited number of existing textbooks. To make the book more accessible to a wide range of readers I have kept the mathematical language at a level suitable for those who have had only an introductory undergraduate course on probability and statistics, and basic courses in calculus and linear algebra. No sacrifice, however, is made to dispense with rigor. In stating theorems I have not always done so under the weakest possible conditions. This allows the reader to readily verify if such conditions are indeed satisfied in most applications given in modern graduate courses without being lost in extra unnecessary mathematical intricacies. The book is not a mere dictionary of mathematical statistical terms. It is also expository in nature, providing examples and putting emphasis on theorems, limit theorems, comparison of different statistical procedures, and statistical distributions. The various topics are covered in appropriate details to give the reader enough confidence in himself (herself) which will then allow him (her) to consult the references given in the Bibliography for proofs, more details, and more applications. At the end of various sections of the book references are given where proofs and/or further details may be found. No attempt is made here to supply historical details on who did what, when. Accordingly, I apologize to any colleague whose name is not found in the list of references or whose name may not appear attached to a theorem or to a statistical procedure. All that should matter to the reader is to obtain quick and precise information on the technical aspects he or she is seeking. To benefit as much as possible from the book, it is advised to consult first the Contents on a given topic, then the Subject Index, and then the section on Notations. Both the Contents and the Subject Index are quite elaborate.

We hope this book will fill a gap, which we feel does exist, and will provide a useful reference to all those concerned with mathematical statistics.

E. B. M.

Contents

Some Notations

$A \subseteq B$: A subset of B and may include equality.

$A \nsubseteq B$: A subset of B and excludes equality.

A^c: complement of A.

$$\binom{n}{x} \equiv \frac{n!}{(n-x)!\,x!}, \qquad n! \equiv \prod_{i=1}^{n} i, \qquad 0! \equiv 1.$$

$I_n = [\delta_{ij}]_{n \times n}$, $(n \times n)$ matrix, and where δ_{ij} is the Kronecker delta, i.e., $\delta_{ij} = 0$ for $i \neq j$, $\delta_{ii} = 1$, $i, j = 1, \ldots, n$.

$c_n = O(n^\delta)$, i.e., $\lim_{n \to \infty} |n^{-\delta} c_n| < \infty$. Sometimes we also write $\lim_{n \to \infty} c_n = O(n^\delta)$.

Gamma function: $\Gamma(z) = \int_0^\infty x^{z-1} e^{-x}\, dx$, Re $z > 0$.

$$\Gamma(n) = (n-1)!, n = 1, 2, \ldots.$$

$\Phi(t)$: Characteristic function.

Standard normal distribution: $\phi(z) = \int_{-\infty}^{z} e^{-x^2/2}\, dx/\sqrt{2\pi}$.

If Z is a random variable having a standard normal distribution, we use the notation: $P[Z \geqslant Z_\alpha] = \alpha$.

If χ^2 is a random variable having a chi-square distribution of v degrees of freedom, we use the notation: $P[\chi^2 \geqslant \chi^2_\alpha(v)] = \alpha$.

If T is a random variable having a Student distribution of v degrees of freedom, we use the notation: $P[T \geqslant T_\alpha(v)] = \alpha$.

If F is a random variable having a Fisher's F-distribution of v_1 and v_2 degrees of freedom, we use the notation: $P[F \geqslant F_\alpha(v_1, v_2)] = \alpha$.

$E[X] = \mu_X$ or just μ (mean); $\sigma^2(X) = E[(X - \mu)^2]$ or just σ^2 (variance); $E[X - \mu)^3]/\sigma^3 = \gamma_1$ (coefficient of skewness); $E[(X - \mu)^4]/\sigma^4 - 3 = \gamma_2$ (kurtosis) adjusted so that $\gamma_2 = 0$ for the normal distribution.

"X has a $N(\mu, \sigma^2)$ distribution" read "X has a normal distribution with mean μ and variance σ^2." "$\mathbf{X}$ has a $N(\boldsymbol{\mu}, \boldsymbol{\Sigma})$ distribution" read "$\mathbf{X}$ has a multivariate normal distribution with mean vector $\boldsymbol{\mu}$ and covariance matrix $\boldsymbol{\Sigma}$."

PART 1

FUNDAMENTALS OF MATHEMATICAL STATISTICS

CHAPTER 1

Basic Definitions, Concepts, Results, and Theorems

§1.1. Probability Concepts

The collection of elements under investigation is called the *population.* An experiment for which the outcome cannot *a priori* be determined, but is known to be one of a set of given possible outcomes, is called a *random experiment.* [Here it is assumed that if the same experiment is repeated any number of times, then the outcome each time is again one of the outcomes in the initial set of possible outcomes.] The set of all possible outcomes of a random experiment is called the *sample space* and will be denoted by $\mathscr{S}$. An element $\omega \in \mathscr{S}$ is called a *sample point.*

A family $\mathscr{A}$ of subsets of $\mathscr{S}$ is called a *sigma-field* if: (i) $A \in \mathscr{A}$, then $A^c \equiv \mathscr{S} - A \in \mathscr{A}$; (ii) $A_1, A_2, \ldots$ are pairwise disjoint sets in $\mathscr{A}$, that is, $A_i \cap A_j = \varnothing$ (null set) for all $i \neq j$, then $\bigcup_{i=1}^{\infty} A_i \in \mathscr{A}$. An element A in $\mathscr{A}$ is called an *event.* An element of the form $\{\omega\}$ is called an *elementary event.*

With each event $A \in \mathscr{A}$, we associate a nonnegative function $P(A)$, called a probability (or probability measure), which satisfies: (i) $0 \leqslant P(A) \leqslant 1$; (ii) $P(\mathscr{S}) = 1$; (iii) for pairwise disjoint events $A_1, A_2, \ldots$ in $\mathscr{A}$, $P(\bigcup_{i=1}^{\infty} A_i) = \sum_{i=1}^{\infty} P(A_i)$. In particular, we note that $\mathscr{S}$ and ϕ are in $\mathscr{A}$. $\mathscr{S}$ is called the *sure event* and $\phi(P(\phi) = 0)$ is called the *impossible event.* The triple $(\mathscr{S}, \mathscr{A}, P)$, consisting of the sample space $\mathscr{S}$, the sigma-field $\mathscr{A}$, and the probability measure P, is called a *probability space.*

A and B are said to be *independent events,* if $P(A \cap B) = P(A)P(B)$. In general, $A_1, \ldots, A_n$ are said to be *mutually independent* if for any subset $\{i_1, \ldots, i_t\}$ of $\{1, \ldots, n\}$ with any $2 \leqslant t \leqslant n$, $P(A_{i_1} \cap \cdots \cap A_{i_t}) = P(A_{i_1}) \ldots P(A_{i_t})$.

The *conditional probability* of A *given* B, is defined by $P(A|B) = P(A \cap B)/P(B)$, $A, B \in \mathscr{A}$ for $P(B) \neq 0$.

$\{A_1, A_2, \ldots\}$ is said to constitute a *complete set of events* if $A_1, A_2, \ldots$ $(\in \mathscr{A})$ are mutually disjoint and $\bigcup_{i=1}^{\infty} A_i = \mathscr{S}$. The set $\{A_1, A_2, \ldots\}$ is also called a partition of the set $\mathscr{S}$.

Poincaré's Additive Theorem. *For any two events* $A_1, A_2 \in \mathscr{A}$, $P(A_1 \cup A_2) = P(A_1) + P(A_2) - P(A_1 \cap A_2)$. *The generalization of this formula for an arbitrary finite number of events is immediate. For example,*

$$P(A_1 \cup A_2 \cup A_3) = \sum_{i=1}^{3} P(A_i) - P(A_1 \cap A_2) - P(A_1 \cap A_3) - P(A_2 \cap A_3) + P(A_1 \cap A_2 \cap A_3).$$

Bonferroni Inequality. *Let* $A_1, \ldots, A_k \in \mathscr{A}$, *then*

$$P(A_1 \cap \cdots \cap A_k) \geqslant 1 - \sum_{i=1}^{k} P(A_i^c).$$

Total Probability Theorem. *Let* $\{A_1, A_2, \ldots\}$ *constitute a complete set of events (a partition of* $\mathscr{S}$*) in* $\mathscr{A}$. *Then for any* $B \in \mathscr{A}$,

$$P(B) = \sum_i P(B|A_i)P(A_i).$$

Bayes' Formula. *For* $P(B) \neq 0$, *we may write*:

$$P(A_j|B) = \frac{P(B|A_j)P(A_j)}{\sum_i P(B|A_i)P(A_i)}.$$

The *Borel field* $\mathscr{B}$ in R (real line) is the smallest sigma-field containing all open intervals $\{x: a < x < b\}$ in R. The Borel field $\mathscr{B}^k$ in R^k (k-dimensional Euclidean space) is the smallest sigma-field containing all open k-rectangles $\{(x_1, \ldots, x_k): a_i < x_i < b_i, i = 1, \ldots, k\}$ in R^k.

A *random variable* X is a function from $\mathscr{S}$ to R such that the set $\{\omega: X(\omega) \in B\}$ is in $\mathscr{A}$ for every $B \in \mathscr{B}$. A vector $\mathbf{X} = (X_1, \ldots, X_k)'$ is called a *random k-vector* if $X_1, \ldots, X_k$ are random variables. [We note, conversely, that if $\{\omega: \mathbf{X}(\omega) \in B\}$ is in $\mathscr{A}$ for every $B \in \mathscr{B}^k$, then $X_1, \ldots, X_k$ are random variables.] The probability of the event $\{\omega: \mathbf{X}(\omega) \in B\}$ will be denoted by $P[\mathbf{X} \in B]$, and is called the *probability distribution* of $\mathbf{X}$.

$\mathbf{X}$ is said to be a *discrete random variable* if there exists a countable set of points $\{\mathbf{x}_1, \mathbf{x}_2, \ldots\}$, $\mathbf{x}_i \in R^k$, $i = 1, 2, \ldots$, such that $\sum_i P[\mathbf{X} = \mathbf{x}_i] = 1$, and $P[\mathbf{X} = \mathbf{x}_i] \equiv f(\mathbf{x}_i)$ is called the *probability mass* function (or just probability mass) of $\mathbf{X}$. In particular, if X is a random variable having values of the form $a + bk$, where a, and $b > 0$ are fixed real numbers, and k runs through the set of values $\{0, 1, 2, \ldots, n\}$ or $\{0, \pm 1, \pm 2, \ldots, \pm n\}$ for n finite *or* infinite, then X is said to have a *lattice* probability *distribution*. For a random vector $\mathbf{X}$ if

$P[\mathbf{X} = \mathbf{x}] = 0$ for every $\mathbf{x} \in R^k$, then $\mathbf{X}$ is called a continuous random k-vector. If, in addition, there exists a nonnegative function f from $(R^k, \mathscr{B}^k)$ to $(R, \mathscr{B})$ such that

$$P[\mathbf{X} \in B] = \int_B f(\mathbf{x})\, d\mathbf{x} \qquad \left(\text{and } P[\mathbf{X} \in R^k] = \int_{R^k} f(\mathbf{x})\, d\mathbf{x} = 1\right),$$

then $\mathbf{X}$ is called an *absolutely continuous random k-vector.* $f(\mathbf{x})$ is called the *probability density* function of $\mathbf{X}$. Throughout the bulk of this work, by a continuous random k-vector it will mean an absolutely continuous random k-vector.

Consider the interval $\{x \leqslant \alpha\}$ in $\mathscr{B}$. Then $P[X \leqslant a]$, denoted by $F(a)$, is called the *cumulative distribution function* (or just the distribution) of the random variable X in question. Similarly, $P[X_1 \leqslant a_1, \ldots, X_k \leqslant a_k] \equiv F(\mathbf{a})$, $\mathbf{a} = (a_1, \ldots, a_k)'$, will be called the *joint distribution* of $X_1, \ldots, X_k$.

Properties of $F(x)$

(i) $0 \leqslant F(x) \leqslant 1$.
(ii) $F(x) \leqslant F(x')$ for $x \leqslant x'$.
(iii) $F(x + \varepsilon) = F(x)$, $\varepsilon \to +0$, that is, $F(x)$ is continuous from the right.
(iv)
$$\lim_{x \to \infty} F(x) \equiv F(+\infty) = 1, \qquad \lim_{x \to -\infty} F(x) = F(-\infty) = 0.$$
(v) If X is a continuous random variable, then
$$\frac{d}{dx} F(x) = f(x), \qquad dF(x) = f(x)\, dx, \qquad F(x) = \int_{-\infty}^{x} f(x')\, dx'.$$
(vi) If X is a discrete random variable taking values $x_1 < x_2 < \cdots$, then $F(x_i) - F(x_{i-1}) = P[X = x_i]$.

Properties of $F(x_1, x_2)$

(i) $0 \leqslant F(x_1, x_2) \leqslant 1$.
(ii) For $x_1 \leqslant x_1'$, $x_2 \leqslant x_2'$, $F(x_1', x_2') + F(x_1, x_2) - F(x_1', x_2) - F(x_1, x_2') \geqslant 0$.
(iii) $F(x_1 + \varepsilon, x_2) = F(x_1, x_2)$, $F(x_1, x_2 + \varepsilon) = F(x_1, x_2)$ for $\varepsilon \to +0$.
(iv) $F(-\infty, x_2) = 0$, $F(x_1, -\infty) = 0$ for all $x_1, x_2 \in R$.
(v) $F(x_1, \infty) \equiv F_{X_1}(x_1)$ is called the *marginal* distribution of X_1, and $F(\infty, x_2) \equiv F_{X_2}(x_2)$ is called the marginal distribution of X_2. X_1 and X_2 are said to be independent if $F(x_1, x_2) = F_{X_1}(x_1) F_{X_2}(x_2)$ for all $x_1, x_2 \in R$, and with an obvious generalization for more than two random variables: $X_1, \ldots, X_n$.
(vi) Let $\mathbf{X} = (X_1, X_2)'$ be a random vector such that X_1 and X_2 are discrete random variables taking, respectively, values $x_{11}, x_{12}, \ldots; x_{21}, x_{22}, \ldots$. We set $P[\{X_1 = x_{1i}\} \cap \{X_2 = x_{2j}\}] = p_{ij}$. Then $\sum_i \sum_j p_{ij} = 1$ and

$p_{i.} = \sum_j p_{ij}$, $p_{.j} = \sum_i p_{ij}$ denote the marginal probability mass functions of X_1 and X_2, respectively. We note that the conditional probability $P[\{X_1 = x_{1i}\}|\{X_2 = x_{2j}\}]$ may be written as $p_{ij}/p_{.j}$. Also formally,

$$E[X_1] = \sum_i x_{1i} p_{i.} \equiv \mu_{X_1}, \qquad \sigma^2(X_1) = \sum_i x_{1i}^2 p_{i.} - \mu_{X_1}^2,$$

$$\mathrm{Cov}(X_1, X_2) = \sum_i \sum_j x_{1i} x_{2j} p_{ij} - \mu_{X_1}\mu_{X_2}.$$

(vii) Let $\mathbf{X} = (X_1, X_2)'$ be a continuous random vector. Then

$$F(a, b) = \int_{-\infty}^{b} \int_{-\infty}^{a} f(x_1, x_2)\, dx_1\, dx_2, \qquad \frac{\partial}{\partial x_1}\frac{\partial}{\partial x_2} F(x_1, x_2) = f(x_1, x_2).$$

Also

$$f_{X_1}(x_1) = \int_{-\infty}^{\infty} f(x_1, x_2)\, dx_2 \quad \text{and} \quad f_{X_2}(x_2) = \int_{-\infty}^{\infty} f(x_1, x_2)\, dx_1,$$

are called the marginal densities of X_1 and X_2. The conditional probability density of X_1 given that $X_2 = x_2$ is defined by $f(x_1|x_2) = f(x_1, x_2)/f_{X_2}(x_2)$ for $f_{X_2}(x_2) \neq 0$. We note that formally,

$$E[X_1] = \int_{-\infty}^{\infty} x_1 f_{X_1}(x_1)\, dx \equiv \mu_{X_1}, \quad \sigma^2(X_1) = \int_{-\infty}^{\infty} x_1^2 f_{X_1}(x_1)\, dx_1 - \mu_{X_1}^2,$$

$$E[X_1|X_2] = \int_{-\infty}^{\infty} x_1 f(x_1|x_2)\, dx_1,$$

$$P[a \leqslant X_1 \leqslant b | X_2 = x_2] = \int_a^b dx_1\, f(x_1|x_2),$$

$$\mathrm{Cov}(X_1, X_2) = \int_{-\infty}^{\infty} \int_{-\infty}^{\infty} x_1 x_2 f(x_1, x_2)\, dx_1\, dx_2 - \mu_{X_1}\mu_{X_2}.$$

Double Expectation. *If* $E[|X|] < \infty$, *then* $E[E[X|Y]] = E[X]$ *for two random variables.*

A number which maximizes the probability density or the probability mass function of a random variable X is called a *mode* of the distribution of X. A number Q_α, called the αth quantile $(0 < \alpha < 1)$ of the distribution of a random variable, is defined by $P[X < Q_\alpha] \leqslant \alpha \leqslant P[X \leqslant Q_\alpha]$. In particular, for $\alpha = 0.5$, $Q_{0.5}$ is called a *median* of the distribution in question.

A random variable X_1 with distribution $F_1(x)$ is said to be *stochastically smaller* than a random variable X_2 with distribution $F_2(x)$ if $F_1(x) \geqslant F_2(x)$ for all x, and with strict inequality holding for at least one x.

Consider two discrete random variables X_1, X_2, if $P[X_1 \neq X_2] = 0$, then we say that $X_1 = X_2$ with *probability one*. Similarly, if $g(x)$ is a continuous function with the exception on a set S, such that $P[X \in S] = 0$, then $g(x)$ is said to be continuous with probability one. Quite generally, if a certain relation holds identically with the exception for points in a set S with probability equal to zero, then the relation is said to hold P-almost everywhere or to hold with probability one.

[Cf., Bickel and Doksum (1977), Billingsley (1979), Burrill (1972), Cramér (1974), Fisz (1980), Fourgeaud and Fuchs (1967), Gnedenko (1966), Milton and Tsokos (1976), Rényi (1970), Roussas (1973), Tucker (1967), Wilks (1962).]

§1.2. Random Samples

Finite Population. Suppose every sample of size n that may selected from the population has the same probability of being selected, then any such a sample is called a *random sample.*

Infinite Population. A set $\{X_1, \ldots, X_n\}$ of independent random variables each having the same population distribution is called a *random sample* of size n. Throughout by a sample we mean a random sample.

A random variable $g(X_1, \ldots, X_n)$ which is a function of independent identically distributed random variables such that if the sample yields $X_1 = x_1, \ldots, X_n = x_n$, then $g(x_1, \ldots, x_n)$ is uniquely determined is called a *statistic.* [We note that a statistic cannot depend on any unknown parameters which may characterize the distribution of the population.]

Let $X_1, \ldots, X_n$ be independent identically distributed random variables. Then Y_k is called the kth *order statistic* of $X_1, \ldots, X_n$ if Y_k is the kth smallest of the $X_1, \ldots, X_n$ observations. We note that $Y_1 \leqslant Y_2 \leqslant \cdots \leqslant Y_n$, where $Y_1 = \min_i X_i$, $Y_n = \max_i X_i$. The random variables $Y_1, \ldots, Y_n$ are called the *order statistics* of the sample $X_1, \ldots, X_n$. $Y_n - Y_1$ is called the *sample range* and Y_1, Y_n are called the *extremes* of the sample. If $X_i \neq X_j$ for all $i \neq j$ in $(1, \ldots, n)$, then the *rank* R_i of X_i is defined by the number $N(i)$ of the X_j less than X_i plus one: $R_i = N(i) + 1$. If $X_{1i}, \ldots, X_{mi}$ denote the X_j equal to X_i (including X_i itself), then the *average rank* or *mid rank* of X_i is defined by $\sum_{j=1}^{m} (N(i) + j)/m$, where $N(i)$ is the number of the X_j less than X_i.

A random variable Q_α is an αth *quantile* of the sample $X_1, \ldots, X_n$, if

$$[\text{number of } X_j < Q_\alpha]/n \leqslant \alpha \leqslant [\text{number of } X_j \leqslant Q_\alpha]/n.$$

In particular, for $\alpha = 0.5$, $Q_{0.5} \equiv M$ is called the *median* of the sample. An equivalent definition for the median is the following: $M = [Y_{n/2} + Y_{(n+2)/2}]/2$ if $n =$ even, and $M = Y_{(n+1)/2}$ if $n =$ odd, where $Y_1, \ldots, Y_n$ are the order statistics of the sample $X_1, \ldots, X_n$.

The function $\hat{F}_n(x) = [\text{number of } X_i \leqslant x]/n$ is called the sample distribution or empirical distribution.

§1.3. Moments

(i) rth *Moment*. $\alpha_r = E[X^r]$, $r > 0$. For $r \geqslant 1$, a positive integer, if α_r exists then so do $\alpha_1, \ldots, \alpha_{r-1}$. For the standard normal distribution $\alpha_{2k} = (2k)!/2^k k!$, $k = 1, 2, \ldots$; $\alpha_{2k+1} = 0$, $k = 0, 1, \ldots$.

(ii) rth *Absolute Moment*. $\beta_r = E[|X|^r]$,

$$(\beta_r)^{1/r} \leqslant (\beta_{r+1})^{1/(r+1)}, \qquad r = 1, 2, \ldots.$$

(iii) rth *Factorial Moment*. $\alpha'_r = E[X(X-1)\ldots(X-r+1)]$.

(iv) rth *Moment About a Point a*. $\mu'_r(a) = E[(X-a)^r]$, and

$$\mu'_r(a) = \sum_{k=1}^{r} \binom{r}{k} \mu'_k(b)(b-a)^{r-k},$$

for two points a and b, and r a natural number.

(v) Let $a = E[X] \equiv \mu$, then we have the moments about the mean:

$\mu_r = E[(X-\mu)^r]$, $\qquad r = 2, 3, \ldots \qquad (\mu_1 \equiv 0)$, μ (mean),

$\mu_2 \equiv \sigma^2$ (variance),

$\mu_3 \equiv \gamma_1 \sigma^3$, γ_1 (coefficient of *skewness*)—a measure of symmetry,

$\mu_4 \equiv (\gamma_2 + 3)\sigma^4$, γ_2 (*kurtosis*)—a measure of peakedness—

adjusted so that $\gamma_2 = 0$ for a normal distribution.

If $\gamma_2 > 0$, the distribution is said to be leptokurtic, and if $\gamma_2 < 0$, the distribution is said to be platykurtic.

The kurtosis γ_2 does not always exist (e.g., for the Cauchy distribution), and the following reasonable measure of *peakedness* has been suggested [Horn (1983)], for distributions with symmetric unimodal densities, which always exists:

$$mt_p(f) = 1 - (p/A_p(f)),$$

where $A_p(f) = f(0) \cdot F^{-1}(p + 0.5)$, for some $0 < p < 0.5$, and without loss of generality the density f has been taken to be symmetric about 0; F^{-1} denotes the inverse of F. The quantity $p/A_p(f)$ denotes the proportion of the area of a rectangle $R_p(f)$ covered by the density, where the rectangle $R_p(f)$, in the $x - y$ plane, is formed by the lines $x = 0$, $y = 0$, $y = f(0)$, $x = F^{-1}(p + 0.5)$, and hence $0 \leqslant mt_p(f) \leqslant 1$. For a rectangular density $mt_p(f) = 0$, and for a density which looks like a spike with a long tail $mt_p(f)$ is close to one. As one is more concerned with the center of the density than with the tail in characterizing its peakedness, one may choose a p-value which is not too large, say $p \leqslant 0.25$. For $p = 0.25$, this measure of peakedness yields to 0.071 for the normal distribution and 0.215 for the Cauchy distribution. This ordering agrees with the intuitive picture that the Cauchy density is more "peaked" than the normal density.

(vi) rth *Cumulant* κ_r is defined upon writing the characteristic function as

$$\Phi(t) = \exp\left[\sum_{r=1}^{\infty} \kappa_r \frac{(it)^r}{r!}\right].$$

Some relations between the moments $\alpha_1, \alpha_2, \ldots$ and the cumulants are:

$$\alpha_1 = \kappa_1, \qquad \alpha_2 = \kappa_2 + \kappa_1^2, \qquad \alpha_3 = \kappa_3 + 3\kappa_2\kappa_1 + \kappa_1^3,$$

$$\alpha_4 = \kappa_4 + 4\kappa_3\kappa_1 + 3\kappa_2^2 + 6\kappa_2\kappa_1^2 + \kappa_1^4.$$

In particular, we note that:

$$\mu = \kappa_1, \qquad \sigma^2 = \kappa_2, \qquad \gamma_1 = \kappa_3/(\kappa_2)^{3/2}, \qquad \gamma_2 = \kappa_4/\kappa_2^2.$$

(vii) The product moments of two random variables X and Y about 0 are defined by $\alpha_{rs} = E[X^r Y^s]$, and about their means μ_X, μ_Y by $\mu_{rs} = E[(X - \mu_X)^r (Y - \mu_Y)^s]$. In particular, the covariance of the two random variables is $\mu_{11} \equiv \sigma_{12} \equiv \mathrm{Cov}(X_1, X_2)$, and the correlation coefficient is $\rho = \mu_{11}/\sqrt{\mu_{20}\mu_{02}}$, $-1 \leqslant \rho \leqslant 1$. If the two random variables X and Y are independent, then $\rho = 0$. However, if $\rho = 0$ this does not necessarily mean that the two random variables X and Y are independent.

(viii) *Moments of the Sample Mean from a Finite Population Without Replacement.* Consider a finite population of size N. Suppose that the probability of choosing any single one of the N elements from the population $\{x_1, \ldots, x_N\}$ is constant. Let $Y_1, \ldots, Y_n$ denote a sample *without* replacement. Clearly, $Y_1, \ldots, Y_n$ are *not* independent. We introduce the population moments:

$$\mu = \sum_{i=1}^{N} x_i/N \equiv \bar{x}, \qquad \mu_r = \sum_{i=1}^{N} (x_1 - \bar{x})^r/N.$$

Then the first four moments of the sample mean $\bar{Y} = \sum_{i=1}^{n} Y_i/n$, with the latter three about the population (sample) mean, are:

$$E[\bar{Y}] = \mu,$$

$$E[(\bar{Y} - \mu)^2] = \frac{N - n}{n(N - 1)} \mu_2,$$

$$E[(\bar{Y} - \mu)^3] = \frac{(N - n)(N - 2n)}{n^2(N - 1)(N - 2)} \mu_3,$$

$$E[(\bar{Y} - \mu)^4] = \frac{N - n}{n^3(N - 1)(N - 2)(N - 3)}\{(N^2 - 6nN + N + 6n^2)\mu_4 + 3N(n - 1)(N - n - 1)\mu_2^2\}.$$

(ix) *Independent Identically Distributed Random Variables.* Let $X_1, \ldots, X_n$ denote n independent identically distributed random variables, and set $\alpha_s = E[X_1^s]$, $s = 1, 2, \ldots$, $\mu_s = E[(X_1 - \alpha_1)^s]$, $s = 2, \ldots$. We define sample moments:

$$a_s = \sum_{i=1}^{n} X_i^s/n, \quad s = 1, 2, \ldots; \quad \text{and} \quad m_s = \sum_{i=1}^{n} (X_i - \bar{X})^s/n, \quad s = 2, 3, \ldots,$$

where $a_1 = \bar{X}$. Then

$$E[a_s] = \alpha_s, \qquad n \geqslant 1,$$

$$E[(a_s - \alpha_s)^2] = \frac{(\alpha_{2s} - \alpha_s^2)}{n}, \qquad n \geqslant 1,$$

$$E[m_s] = \mu + \frac{\frac{1}{2}s(s - 1)\mu_{s-2}\mu_2 - \mu_s}{n} + O\left(\frac{1}{n^2}\right), \qquad n \to \infty, \quad s \neq 1,$$

$$E[(m_s - \mu_s)^2] = \frac{\mu_{2s} - \mu_s^2 - 2s\mu_{s-1}\mu_{s+1} + s^2\mu_2\mu_{s-1}^2}{n} + O\left(\frac{1}{n^2}\right),$$

$$n \to \infty, \quad s \neq 1.$$

Finally for the moments $m_s' = \sum_{i=1}^{n} (X_i - \alpha_1)^s/n$, we have

$$E[m_s'] = \mu_s, \qquad s \neq 1, \quad n \geqslant 1.$$

$$E[(m_s' - \mu_s)^2] = \frac{\mu_{2s} - \mu_s^2}{n}, \qquad s \neq 1, \quad n \geqslant 1.$$

[For $s = 1$ simply set $\mu_s = 0$ in these latter two equations.]

(See also §2.15.)

[Cf., Cramér (1974), Kendall and Stuart (1977), Serfling (1980), Horn (1983) Whetherill (1981).]

§1.4. Some Inequalities Involving Probabilities and Moments

(i) Let X and Y be random variables such that $E[Y^2] < \infty$. Then for any function $h(x)$ such that $E[h^2(X)] < \infty$,

$$E[(Y - h(X))^2] \geqslant E[(Y - E[Y|X])^2],$$

with strict inequality holding unless $h(x) = E[Y/x]$.

(ii) In (i) if $h(x) = E[Y]$ for all x,

$$\sigma^2(Y) \geqslant E[(E[Y] - E[Y|X])^2],$$

with strict inequality holding unless Y is a function of X given by $Y = E[Y/X]$.

(iii) Let X be a random variable and $h(x)$ be a nonnegative function such that $E[h(X)] < \infty$, then for any $\varepsilon > 0$,

$$P[h(X) \geqslant \varepsilon] \leqslant E[h(X)]/\varepsilon.$$

(iv) *Chebyshev Inequality.* In (iii) if $h(x) = x^2$, by noting that $P[|X| \geqslant \varepsilon] = P[X^2 \geqslant \varepsilon^2]$, we have

$$P[|X| \geqslant \varepsilon] \leqslant E[X^2]/\varepsilon^2, \qquad \varepsilon > 0.$$

(v) *Markov Inequality.* In (iii) if $h(x) = |x|^r$, for $r > 0$, by noting that $P[|X| \geqslant \varepsilon] = P[|X|^r \geqslant \varepsilon^r]$, we have

$$P[|X| \geqslant \varepsilon] \leqslant E[|X|^r]/\varepsilon^r, \qquad \varepsilon > 0.$$

(vi) *Holder Inequality.* Let X and Y be two random variables and $p > 0$, $q > 0$ two numbers such that $p^{-1} + q^{-1} = 1$, and $E[|X|^p] < \infty$, $E[|Y|^q] < \infty$, then

$$E[|XY|] \leqslant (E[|X|^p])^{1/p}(E[|Y|^q])^{1/q}.$$

(vii) *Schwarz Inequality.* In (vi) if $p = q = \frac{1}{2}$, we have

$$E[|XY|] \leqslant \sqrt{E[X^2]E[Y^2]}.$$

(viii) *Jensen Inequality.* Let X be a random variable, and $h(x)$ be a *convex* function, i.e., for any numbers $\alpha_1, \alpha_2, 0 \leqslant \varepsilon \leqslant 1$,

$$h(\varepsilon\alpha_1 + (1 - \varepsilon)\alpha_2) \leqslant \varepsilon h(\alpha_1) + (1 - \varepsilon)h(\alpha_2).$$

If $E[h(X)]$ exists then $h[E(X)] \leqslant E[h(X)]$. $h(x)$ is called a *concave* function if we have instead

$$h(\varepsilon\alpha_1 + (1 - \varepsilon)\alpha_2) \geqslant \varepsilon h(\alpha_1) + (1 - \varepsilon)h(\alpha_2),$$

and we then have $h[E(X)] \geqslant E[h(X)]$.

(ix) *Berge Inequality.* Let $(X, Y)'$ be a random 2-vector such that $E[X] = 0$, $E[Y] = 0$, $\sigma^2(X) = 1$, $\sigma^2(Y) = 1$, $\rho = \mathrm{Cov}(X, Y)$. Then for any $\varepsilon > 0$,

$$P[\max(|X|, |Y|) \geqslant \varepsilon] \leqslant (1 + \sqrt{1 - \rho^2})/\varepsilon^2.$$

(x) *Kolmogorov Inequality*. Let $X_1, \ldots, X_n$ be independent random variables such that $E[X_i] = 0$ for $i = 1, \ldots, n$, and $\sigma^2(X_i) \equiv \sigma_i^2 < \infty$. Then for any $\varepsilon > 0$,

$$P[\max(|X_1|, |X_1 + X_2|, \ldots, |X_1 + \cdots + X_n|) \geqslant \varepsilon] \leqslant \sum_{i=1}^{n} \sigma_i^2/\varepsilon^2.$$

(xi) *Šidák Inequality*. Let $\mathbf{X} = (X_1, \ldots, X_k)'$ be a random k-vector having a $N(\mathbf{0}, \boldsymbol{\Sigma})$ distribution. Let vS^2 be an independent random variable from $\mathbf{X}$ having a chi-square distribution with v degrees of freedom, then for any $\varepsilon_1 > 0, \ldots, \varepsilon_k > 0$,

$$E[|X_1| \leqslant \varepsilon_1 S, \ldots, |X_k| \leqslant \varepsilon_k S] \geqslant \prod_{i=1}^{k} P[|X_i| \leqslant \varepsilon_i S].$$

[Cf., Cramér (1974), Roussas (1973), Bickel and Doksum (1977), Rao (1973), Miller (1981).]

§1.5. Characteristic Functions

(i) The characteristic function of a random variable X is defined by $\Phi(t) = E[\exp(itX)]$, t real. $\Phi(0) = 1$, $|\Phi(t)| \leqslant 1$. Suppose that $E[|X|^n] < \infty$ for some positive integer n, then

$$\Phi(t) = \sum_{k=0}^{n} \alpha_k \frac{(it)^k}{k!} + R_n(t), \quad \text{where } \alpha_k = E[X^k],$$

$$R_n(t) = \frac{[\Phi^{(n)}(\theta t) - (i)^n \alpha_n] t^n}{n!}, \quad \text{for some } 0 < \theta < 1,$$

$$|R_n(t)| \leqslant \frac{2|t|^n E[|X|^n]}{n!}, \quad \text{and} \quad (d/dt)^k \Phi(t) \equiv \Phi^{(k)}(t),$$

$$\Phi^{(k)}(0) = (i)^k \alpha_k, \qquad k = 0, 1, \ldots, n.$$

For any real constants a and b, the characteristic function $ax + b$ is given by $e^{itb}\Phi(at)$, where $\Phi(t)$ is the characteristic function of X.

(ii) *Inversion Formulas I*. Suppose x_1 and x_2 are two continuity points of the distribution $F(x)$ of a random variable X with characteristic function $\Phi(t)$, then

$$F(x_2) - F(x_1) = \frac{1}{2\pi} \int_{-\infty}^{\infty} \frac{e^{-ix_1 t} - e^{-ix_2 t}}{it} \Phi(t)\, dt,$$

and F is uniquely determined by $\Phi(t)$.

Multivariate Generalization. Let $\Phi(\mathbf{t}) = E[\exp(i\mathbf{t}'\mathbf{X})]$ denote the characteristic function of a random k-vector $\mathbf{X} = (X_1, \ldots, X_k)'$ with distribution $F(\mathbf{x})$. Let $\mathbf{a} = (a_1, \ldots, a_k)'$ and $\mathbf{b} = (b_1, \ldots, b_k)'$ be continuity points of F, with $a_i \leqslant b_i$, $i = 1, \ldots, k$, then

$$F(\mathbf{b}) - F(\mathbf{a}) = \frac{1}{(2\pi)^k} \lim_{\begin{pmatrix} L_1 \to \infty \\ \vdots \\ L_k \to \infty \end{pmatrix}} \int_{-L_k}^{L_k} \cdots \int_{-L_1}^{L_1}$$

$$\times \left[\prod_{j=1}^{k} \left(\frac{e^{-ia_jt} - e^{-ib_jt}}{it} \right) \right] \Phi(\mathbf{t})\, dt_1, \ldots dt_k,$$

and F is uniquely determined by Φ.

(iii) *Inversion Formulas II.* Let X be a random variable with characteristic function $\Phi(t)$. If $\int_{-\infty}^{\infty} |\Phi(t)|\, dt < \infty$, then $F'(x) = f(x)$ exists and is a bounded continuous function and

$$f(x) = \frac{1}{2\pi} \int_{-\infty}^{\infty} e^{-itx}\Phi(t)\, dt, \qquad \Phi(t) = \int_{-\infty}^{\infty} e^{itx} f(x)\, dx.$$

More generally, if X is a continuous random variable then

$$f(x) = \lim_{\delta \to \infty} \lim_{L \to \infty} \frac{1}{2\pi} \int_{-L}^{L} \frac{1 - e^{-it\delta}}{it\delta} e^{-itx}\Phi(t)\, dt.$$

If X is a discrete random variable then

$$f(x) = \lim_{L \to \infty} \frac{1}{2L} \int_{-L}^{L} e^{-itx}\Phi(t)\, dt.$$

In particular, if X has a lattice distribution (see §1.1), then $\Phi(t + 2\pi) = \Phi(t)$ and

$$f(x) = \frac{1}{2\pi} \int_{-\pi}^{\pi} e^{-itx}\Phi(t)\, dt.$$

(iv) *Uniqueness of Characteristic Functions.* Two distributions with identical characteristic functions are equal.

(v) *Independence of Random Variables* (Theorem). Let $X_1, \ldots, X_k$ be random variables with characteristic functions $\Phi_1(t), \ldots, \Phi_k(t)$, respectively. Let $\Phi(\mathbf{t})$ denote the characteristic function of the random vector $\mathbf{X} = (X_1, \ldots, X_k)'$. Then the random variables $X_1, \ldots, X_k$ are independent if and only if $\Phi(\mathbf{t}) = \Phi_1(t_1) \ldots \Phi_k(t_k)$ for all $-\infty < t_i < \infty$, $i = 1, \ldots, k$, where $\mathbf{t} = (t_1, \ldots, t_k)'$. For several examples of characteristic functions associated with various distributions see Chapter 3.

[Cf., Cramér (1974), Manoukian (1986), Burrill (1972).]

§1.6. Moment Generating Functions

A random variable X has a moment generating function, if there is a positive number $d > 0$, such that $E[e^{tX}] < \infty$ for all $-d < t < d$, and it is then defined by $M(t) = E[e^{tX}]$, for $-d < t < d$. To have an infinite series expansion

$$M(t) = \sum_{k=0}^{\infty} \alpha_k \frac{(t)^k}{k!}, \quad \text{where} \quad \alpha_k = E[X^k], \qquad k = 1, 2, \ldots,$$

are the moments, it is sufficient [e.g., Kirmani and Isfahani (1983)] to choose $-(d/2) < t < (d/2)$. The latter means that for $-(d/2) < t < (d/2)$,

$$\lim_{N\to\infty} \left| \sum_{k=1}^{N} \alpha_k \frac{(t)^k}{k!} - M(t) \right| = 0.$$

§1.7. Determination of a Distribution from Its Moments

Let α_r denote the rth moment of a random variable. If the series $\sum_{r=1}^{\infty} (\alpha_r/r!)(\delta)^r$ converges in some interval $0 < \delta < s$, then the moments α_1, $\alpha_2, \ldots$ determine the distribution $F(x)$ of the random variable uniquely.

[Cf., Billingsley (1979), Manoukian (1986).]

§1.8. Probability Integral Transform

(i) Let X be a continuous random variable with distribution $F(x)$, then the random variable $F(X)$ is uniformly distributed on (0, 1) and is called the probability integral transform of X.

(ii) *The David–Johnson Problem.* For the case when $F(x)$ depends on unknown parameters estimated by a sample see §2.18.

[Cf., Manoukian (1986).]

§1.9. Unbiased and Asymptotically Unbiased Estimators

Let $X_1, \ldots, X_n$ be independent identically distributed random variables each with a distribution F_θ depending on a parameter $\theta \in \Omega$, where Ω is called the parameter space. If $T_n(X_1, \ldots, X_n)$ is a statistic such that $E_\theta[T_n] = g(\theta)$, for all n and for all $\theta \in \Omega$, then T_n is called an unbiased estimator of $g(\theta)$. If $E_\theta[T_n] - \theta \equiv b_n(\theta) \neq 0$, then $b_n(\theta)$ is called the bias of the estimator of $T_n(\theta)$ of θ, where θ is the true value of the parameter in question. Suppose there exist functions $\psi_n(\theta)$, $\sigma_n^2(\theta)$, such that $(T_n - \psi_n(\theta))/\sigma_n(\theta)$ has, for $n \to \infty$, some limiting distribution with mean zero and variance 1 (such as the $N(0, 1)$ distribution) for all $\theta \in \Omega$, then T_n is called an asymptotically unbiased estimator of a function $g(\theta)$ if $\lim_{n\to\infty}(g(\theta) - \psi_n(\theta))/\sigma_n(\theta) = 0$. [$\psi_n(\theta)$ and $\sigma_n^2(\theta)$ are referred to as the asymptotic mean and variance of T_n, respectively.]

§1.10. Uniformly Minimum Variance Unbiased Estimators

An unbiased estimator of a parameter $\theta \in \Omega$ which is of minimum variance, for all $\theta \in \Omega$, in the class of all unbiased estimators of $\theta \in \Omega$, is called a uniformly minimum variance unbiased (UMVU) estimator of θ.

§1.11. Consistency of an Estimator

Let $X_1, \ldots, X_n$ be independent identically distributed random variables each with a distribution F_θ depending on a parameter θ. A statistic $T_n(X_1, \ldots, X_n) \equiv T_n(\mathbf{X})$ is called a (weakly) consistent estimator of a function $g(\theta)$, if T_n converges, for $n \to \infty$, in probability to $g(\theta)$. By consistency of an estimator one normally refers to weak consistency. T_n is said to be a strongly consistent estimator of $g(\theta)$ if it converges, for $n \to \infty$, with probability one to $g(\theta)$.

[For other and more general definitions see, for example, Serfling (1980).]

§1.12. M-Estimators

Let $X_1, \ldots, X_n$ denote a sample, that is independent identically distributed random variables. An estimate of a parameter θ obtained by minimizing, in general, an expression like $\sum_{i=1}^n \rho(X_i, \theta)$, with respect to θ, is called an *M-estimator.*

This includes a very large class of estimators. For example, by choosing $\rho(x_i, \theta) = -\ln f(x_i; \theta)$ where $f(x; \theta)$ is the probability density or the probability mass in question, we obtain the maximum likelihood estimator (see §2.24). As another example, suppose that each X_i has a continuous distribution $F(x - \theta)$, where F is symmetric about the origin ($F(x) = 1 - F(-x)$). We may then take $\rho(x_i, \theta)$ as a measure of distance between x_i and the location parameter θ, $\rho(x_i, \theta) \equiv \rho(x_i - \theta)$. Typically, one may take $\rho(x) = x^2$ and this will lead to the *least-squares estimate* of θ to be the sample mean. More generally, we may choose (Huber) $\rho_H(x) = \frac{1}{2}x^2$ for $-k \leqslant x \leqslant k$, and $\rho_H(x) = k|x| - \frac{1}{2}k^2$, for $x > k$, $x < -k$, and k is some positive number. Such a definition has an advantage over the x^2 form as it puts less emphasis on extreme observations, and by choosing k large enough, it reduces to the x^2 form. The function $\rho(x)$ is called the *objective* function. More is said about M-estimators in §1.42 where suggestions are made on how one may choose an objective function. See also §§2.24 and 2.25 for various asymptotic theorems of M-estimators.

§1.13. *L*-Estimators and the α-Trimmed Mean

A statistic which is a linear combination of order statistics (see §1.2) is called an "*L-estimator.*" Generally, a statistic which is a linear combination of functions of order statistics: $\sum a_i h(X_{(i)})$, is also called an "*L-estimator.*"

The α-trimmed mean, for example, is an "*L*-estimator" and is defined by:

$$\bar{X}_\alpha = (n - 2[n\alpha])^{-1} \sum_{i=[n\alpha]+1}^{n-[n\alpha]} X_{(i)}, \qquad 0 < \alpha < \tfrac{1}{2},$$

where $X_{(1)} \leqslant \cdots \leqslant X_{(n)}$ are the order statistics of independent identically distributed random variables $X_1, \ldots, X_n$, and $[x]$ is the largest positive integer $\leqslant x$. For $\alpha \to 0$, this reduces to the sample mean $\bar{X}$, and for $\alpha \to \frac{1}{2}$, this reduces to the sample median.

As an estimator of a location parameter of a distribution, how much trimming should be done? That is, what is a suitable value for α? For a specific distribution, the variances $\sigma^2(\bar{X}_\alpha)$, for different α, may be, in principle, determined and an efficiency comparison then may be made. A detailed analysis of such efficiency comparisons [Rosenberger, J. L. and Gasko, M. *in* Hoaglin *et al.* (1983)] has been carried out for a broad class *C* of symmetric distributions, from the light-tailed (such as the Gaussian) to the heavy-tailed (such as the slash), which include the Gaussian, the logistic and the slash distributions suggests the following trimmings. If the distribution of the underlying population is unknown but is known to belong to the broad class *C* from the light-tailed to the heavy-tailed symmetric distributions, then a 25% trimming (corresponding to the so-called midmean) is suggested. When heavy-tailed distributions are anticipated and the sample size $n \leqslant 20$, then slightly more than 25% trimming is recommended. When tails as heavy as those of Cauchy and slash are not reasonable a 20% trimming is suggested.

See also §§1.42 and 2.19.

§1.14. *R*-Estimators

A statistic which is a linear combination of (or functions of) rank statistics (see §1.2): $\sum_{i=1}^{n} a_i h(R_i)$, is called an "*R-estimator.*" The constants $h(1), \ldots, h(n)$ are called *scores*, and $a_1, \ldots, a_n$ are called *regression constants.* Such statistics are referred to as linear rank statistics.

For some examples of *R*-estimators (statistics) see §1.44.

§1.15. Hodges–Lehmann Estimator

One-Sample Problem. Let $X_1, \ldots, X_n$ be a sample from a continuous distribution $F(x - \theta)$, where F is *symmetric* about the origin ($F(x) = 1 - F(-x)$). Consider the test H_0: $\theta = 0$ against H_A: $\theta > 0$, and suppose that H_0 is rejected for $T_n(X_1, \ldots, X_n) \geqslant t_n$, where $T_n(X_1, \ldots, X_n)$ is the statistic in question. Suppose

(a) $T_n(x_1 + a, \ldots, x_n + a) \geqslant T_n(x_1, \ldots, x_n)$ for all $a \geqslant 0$ and every $x_1, \ldots, x_n$.
(b) When H_0 is true, $T_n(X_1, \ldots, X_n)$ is symmetrically distributed about some point ξ_n.

Then the Hodges–Lehmann estimator $\hat{\theta}$ of θ is defined by:

$$\hat{\theta} = (\bar{\theta} + \underline{\theta})/2,$$

where

$$\bar{\theta} = \sup\{\theta: T_n(X_1 - \theta, \ldots, X_n - \theta) > \xi_n\},$$

and

$$\underline{\theta} = \inf\{\theta: T_n(X_1 - \theta, \ldots, X_n - \theta) < \xi_n\}.$$

Theorem. *If, in addition, $T_n(x_1, \ldots, x_n) - \xi_n = \xi_n - T_n(-x_1, \ldots, -x_n)$ for every $x_1, \ldots, x_n$, then $\hat{\theta} = \hat{\theta}(X_1, \ldots, X_n)$ is symmetrically distributed about θ and is an unbiased (provided it exists) estimator of θ.*

For example, the Hodges–Lehmann estimators in the sign test and the Wilcoxon signed rank test (see §1.44) are, respectively, $\hat{\theta} = \text{median}\{X_i, 1 \leqslant i \leqslant n\}$, $\hat{\theta} = \text{median}\{(X_i + X_j)/2, 1 \leqslant i \leqslant j \leqslant n\}$, and are both unbiased estimators of θ (if $F(x)$ is symmetric about the origin $F(x) = 1 - F(-x)$): $E[\theta] = \theta$, where θ is the median.

Two-Sample Problem. Let $X_{11}, \ldots, X_{1n_1}$, and $X_{21}, \ldots, X_{2n_2}$ be two independent samples from continuous distributions $F(x)$ and $F(x - \theta)$, respectively. Consider the test H_0: $\theta = 0$ against H_A: $\theta > 0$ and suppose H_0 is rejected for $T(X_{11}, \ldots, X_{1n_1}; X_{21}, \ldots, X_{2n_2}) \geqslant t_{n_1,n_2}$ where T is the statistic in question. Suppose

(a) $T(x_{11}, \ldots, x_{1n_1}; x_{21} + a, \ldots, x_{2n_2} + a) \geqslant T(x_{11}, \ldots, x_{1n_1}; x_{21}, \ldots, x_{2n_2})$ for all $a \geqslant 0$ and every $x_{11}, \ldots, x_{1n_1}, x_{21}, \ldots, x_{2n_2}$.
(b) When H_0 is true, $T(X_{11}, \ldots, X_{1n_1}; X_{21}, \ldots, X_{2n_2})$ is symmetrically distributed about some point ξ_{n_1,n_2}.

Then the Hodges–Lehmann estimator $\hat{\theta}$ of θ is defined by

$$\hat{\theta} = (\bar{\theta} + \underline{\theta})/2,$$

where

$$\bar{\theta} = \sup\{\theta: T(X_{11}, \ldots, X_{1n_1}; X_{21} - \theta, \ldots, X_{2n_2} - \theta) > \xi_{n_1,n_2}\},$$

and

$$\underline{\theta} = \inf\{\theta: T(X_{11}, \ldots, X_{1n_1}; X_{21} - \theta, \ldots, X_{2n_2} - \theta) < \xi_{n_1,n_2}\}.$$

Theorem. *If, in addition,*

$$T(x_{11}, \ldots, x_{1n_1}; x_{21}, \ldots, x_{2n_2}) - \xi_{n_1,n_2}$$
$$= \xi_{n_1,n_2} - T(-x_{11}, \ldots, -x_{1n_1}; -x_{21}, \ldots, -x_{2n_2}),$$
$$T(x_{11} + h, \ldots, x_{1n_1} + h; x_{21} + h, \ldots, x_{2n_2} + h)$$
$$= T(x_{11}, \ldots, x_{1n_1}; x_{21}, \ldots, x_{2n_2}),$$

for every $h, x_{11}, \ldots, x_{1n_1}, x_{21}, \ldots, x_{2n_2}$, *and* $F(x)$ *is symmetric about some point (such as the origin), then* $\hat{\theta}$ *is symmetrically distributed about* θ *and is an unbiased estimator of* θ.

For example, for the Wilcoxon–Mann–Whitney test (see §1.44) $\hat{\theta} = \text{median}\{X_{2i} - X_{1j}, j = 1, \ldots, n_1, i = 1, \ldots, n_2\}$ and is an unbiased estimator (if $F(x)$ is symmetric about the origin) of θ, where θ is the median.

See also §§1.44 and 2.29.

[Cf., Randles and Wolfe (1979).]

§1.16. *U*-Statistics

One-Sample U-Statistics. Let X_1, X_2, ... be independent identically distributed random variables each with a distribution $F(x)$. Suppose there is a parameter $\theta(F)$ for which there exists an unbiased estimator $h(X_1, \ldots, X_m)$: $E_F[h(X_1, \ldots, X_m)] = \theta(F)$, where (without loss of generality) h is assumed to be symmetric, that is, it is invariant under permutations of X_1, ..., X_m. The function h is called a kernel of the parameter $\theta(F)$. The *U*-statistic associated with h and $\theta(F)$, and based on a sample $X_1, \ldots, X_n$ of size $n \geqslant m$ is defined by

$$U(X_1, \ldots, X_n) = \frac{1}{\binom{n}{m}} \sum_c h(X_{i_1}, \ldots, X_{i_m}),$$

where the summation is over all $\binom{n}{m}$ combinations of m distinct elements $\{i_1, \ldots, i_m\}$ from the set $\{1, \ldots, n\}$. Clearly, U is an unbiased estimator of $\theta(F)$. The variance of U is given by

$$\text{Var}_F(U) = \frac{1}{\binom{n}{m}} \sum_{i=1}^{m} \binom{m}{i}\binom{n-m}{m-i} \xi_i,$$

$$\xi_i = E_F[h(X_1, \ldots, X_m)h(X_1, \ldots, X_i, X_{m+1}, \ldots, X_{2m-i})] - (\theta(F))^2$$

assuming the latter exist, where $i = 1, \ldots, m$.

As an example, consider the mean $\theta(F) = E_F[X]$, with the kernel $h(x) = x$.

The associated U-statistic based on a sample $X_1, \ldots, X_n$

$$U(X_1, \ldots, X_n) = \frac{1}{\binom{n}{1}} \sum_{i=1}^{n} X_i = \bar{X}$$

is the sample mean, and $\sigma^2(U) = \sigma^2(X)/n$, assuming $\sigma^2(X) < \infty$.

As another example, let $\theta = \sigma^2$ and define the kernel $h(x_1, x_2) = (x_1 - x_2)^2/2$, $m = 2$. Then

$$E[h(X_1, X_2)] = \sigma^2 \quad \text{and} \quad U(X_1, \ldots, X_n) = \sum_{i=1}^{n} (X_i - \bar{X})^2/(n-1).$$

Corresponding to the *sign test* (see §1.44) define $h(x) = \psi(x)$, $m = 1$, where $\psi(x) = 1$ if $x > 0$ and $\psi(x) = 0$ otherwise. Then set $E[h(X)] = p$, and we have:

$$U(X_1, \ldots, X_n) = \sum_{i=1}^{n} \psi(X_i)/n = B/n.$$

Corresponding to the Wilcoxon signed rank test (see §1.44), let $h_1(x) = \psi(x)$, $h_2(x_1, x_2) = \psi(x_1 + x_2)$, where $\psi(x) = 1$ for $x > 0$ and $\psi(x) = 0$ otherwise. Define

$$U_1 = \sum_{i=1}^{n} \psi(X_i)/n, \qquad U_2 = \frac{1}{\binom{n}{2}} \sum_{1 \leqslant i < j \leqslant n} \psi(X_i + X_j).$$

Then $W = nU_1 + \binom{n}{2} U_2$.

Two-Sample U-Statistics. Consider two independent collections $\{X_{11}, X_{12}, \ldots\}$ and $\{X_{21}, X_{22}, \ldots\}$ of independent random variables with distributions $F(x)$ and $G(x)$, respectively. Suppose there is a parameter $\theta = \theta(F, G)$ for which there exists an unbiased estimator $h(X_{11}, \ldots, X_{1m_1}; X_{21}, \ldots, X_{2m_2})$:

$$E[h(X_{11}, \ldots, X_{1m_1}; X_{21}, \ldots, X_{2m_2})] = \theta,$$

where (without loss of generality) it is assumed that h is symmetric within each of its two sets of arguments. The U-statistic associated with h and θ, and based on two samples $X_{11}, \ldots, X_{1n_1}; X_{21}, \ldots, X_{2n_2}$, of sizes $n_1 \geqslant m_1$, $n_2 \geqslant m_2$, is defined by

$$U(X_{11}, \ldots, X_{1n_1}; X_{21}, \ldots, X_{2n_2}) = \frac{1}{\binom{n_1}{m_1}\binom{n_2}{m_2}} \sum_{c} h(X_{1i_1}, \ldots, X_{1i_{m_1}}; X_{2j_1}, \ldots, X_{2j_{m_2}}),$$

where the sum is over all $\binom{n_1}{m_1}\binom{n_2}{m_2}$ combinations of m_1 distinct elements $\{i_1, \ldots, i_{m_1}\}$ from $\{1, \ldots, n_1\}$ and m_2 distinct elements $\{j_1, \ldots, j_{m_2}\}$ from $\{1, \ldots, n_2\}$. U is an unbiased estimator of θ, and its variance is

$$\operatorname{Var} U = \frac{1}{\binom{n_1}{m_1}\binom{n_2}{m_2}} \sum_{i=0}^{m_1} \sum_{j=0}^{m_2} \binom{m_1}{i}\binom{n_1 - m_1}{m_1 - i}\binom{m_2}{j}\binom{n_2 - m_2}{m_2 - j} \xi_{i,j},$$

$$\begin{aligned}\xi_{i,j} = E[h(&X_{11}, \ldots, X_{1m_1}; X_{21}, \ldots, X_{2m_2}) \\ &\times h(X_{11}, \ldots, X_{1i}, X_{1m_1+1}, \ldots, X_{12m_1-i}; \\ &\quad X_{21}, \ldots, X_{2j}, X_{2m_2+1}, \ldots, X_{22m_2-j})] - \theta^2, \qquad \xi_{0,0} \equiv 0,\end{aligned}$$

assuming the latter exist, where $0 \leqslant i \leqslant m_1, 0 \leqslant j \leqslant m_2$.

As an example, consider $\theta = \theta(F, G) = P[X_1 \leqslant X_2]$, with $h(x_1, x_2) = \psi(x_2 - x_1)$, ($\psi(t) = 1$ for $t > 0$, $\psi(t) = 0$ for $t \leqslant 0$), then the associated U-statistic is

$$U(X_{11}, \ldots, X_{1n_1}; X_{21}, \ldots, X_{2n_2}) = \frac{1}{n_1 n_2} \sum_{i=1}^{n_1} \sum_{j=1}^{n_2} \psi(X_{2j} - X_{1i})$$

and is (up to the $1/n_1 n_2$ factor) the so-called Mann–Whitney statistic (see §1.44). In the notation of §1.44, $E[h(X_{11}, X_{12})] = p_1$.

k-Sample U-Statistics. Consider k independent collections $\{X_{11}, X_{12}, \ldots\}$, $\ldots$, $\{X_{k1}, X_{k2}, \ldots\}$ of independent random variables with distributions $F_1(x)$, $\ldots$, $F_k(x)$, respectively. Suppose there is a parameter $\theta = \theta(F_1, \ldots, F_k)$ for which there exists an unbiased estimator $h(X_{11}, \ldots, X_{1m_1}; \ldots; X_{k1}, \ldots, X_{km_k})$: $E[h(X_{11}, \ldots, X_{1m_1}; \ldots; X_{k1}, \ldots, X_{km_k})] = \theta$, where (without loss of generality) it is assumed that h is symmetric within each of its k sets of arguments. The associated U-statistic, based on samples $(X_{11}, \ldots, X_{1n_1})$, $\ldots$, $(X_{k1}, \ldots, X_{kn_k})$, is defined by

$$U(X_{11}, \ldots, X_{1n_1}; \ldots; X_{k1}, \ldots, X_{kn_k}) = \frac{1}{\binom{n_1}{m_1}\cdots\binom{n_k}{m_k}} \sum_c h(X_{1i_{11}}, \ldots, X_{1i_{1m_1}}; \ldots; X_{ki_{k1}}, \ldots, X_{ki_{km_k}}),$$

where the sum is over all $\binom{n_1}{m_1}\cdots\binom{n_k}{m_k}$ combinations of m_1 distinct elements $\{i_{11}, \ldots, i_{1m_1}\}$ from $\{1, \ldots, n_1\}, \ldots,$ of m_k distinct elements $\{i_{k1}, \ldots, i_{km_k}\}$ from $\{1, \ldots, n_k\}$. U is an unbiased estimator of θ. The variance of U may be written down by inspection from the two-sample problem and its expression is to cumbersome to give here.

As an example, consider $\theta = \sum_{1 \leqslant i < j \leqslant k} P[X_i < X_j]$, $X_1, \ldots, X_k$ with continuous distributions $F_1(x), \ldots, F_k(x)$, and kernel

$$h(x_{11}; x_{21}; \ldots; x_{k1}) = \sum_{1 \leqslant i < j \leqslant k} \psi(x_{j1} - x_{i1}),$$

$n_1 = \cdots = n_k \equiv n$, then the U-statistic is

$$U(X_{11}, \ldots, X_{1n}; \ldots; X_{k1}, \ldots, X_{kn}) = \frac{1}{n^2} \sum_{1 \leqslant i < j \leqslant k} \left[\sum_{\alpha=1}^{n} \sum_{\beta=1}^{n} \psi(X_{j\beta} - X_{i\alpha}) \right]$$

and is a so-called Terpstra–Jonckheere type.

See also §§1.44 and 2.30.

[Cf., Randles and Wolfe (1979).]

§1.17. Cramér–Rao–Frêchet Lower Bound

(i) Let $X_1, \ldots, X_n$ be independent identically distributed random variables each with probability density or probability mass function $f(x; \theta)$. Let $T(X_1, \ldots, X_n) \equiv T(\mathbf{X})$ be any statistic such that $\sigma^2(T(\mathbf{X})) < \infty$. Denote $E_\theta[T(\mathbf{X})]$ by $g(\theta)$. Suppose

(a) The set S of positivity of $L(\mathbf{x}; \theta) = \prod_{i=1}^{n} f(x_i; \theta)$ is independent of θ.

(b) For all θ in one or more open intervals, $(\partial/\partial\theta)L(\mathbf{x}; \theta)$ exists for all $\mathbf{x}$ in S and is continuous in θ, (we denote the union of these open intervals by Ω), and $|(\partial/\partial\theta)L(\mathbf{x}; \theta)| \leqslant h(\mathbf{x})$, such that

$$\int_{-\infty}^{\infty} \int_{-\infty}^{\infty} |T(\mathbf{x})| h(\mathbf{x})\, dx_1 \ldots dx_n < \infty, \quad \int_{-\infty}^{\infty} \int_{-\infty}^{\infty} h(\mathbf{x})\, dx_1 \ldots dx_n < \infty.$$

(For the discrete case replace the integrals by summation signs.)

(c) Suppose $0 < I(\theta) < \infty$, where

$$I(\theta) = nE_\theta[((\partial/\partial\theta) \ln f(X; \theta))^2].$$

Then $g'(\theta) = (d/d\theta)g(\theta)$ exists and $\sigma^2(T) \geqslant (g'(\theta))^2/I(\theta)$. If $T(\mathbf{X})$ is an unbiased estimator of θ, that is $g(\theta) = \theta$, then $\sigma^2(T) \geqslant 1/I(\theta)$. The number $1/I(\theta)$ is often referred to as the Cramér–Rao–Frêchet lower bound. The expression $E_\theta[((\partial/\partial\theta) \ln f(X; \theta))^2]$ is called *Fisher's information* about the parameter θ.

(ii) If $T(\mathbf{X})$ in (i) is such that $\sigma^2(T) = (g'(\theta))^2/I(\theta)$ for all $\theta \in \Omega$, that is the Cramér–Rao–Frêchet lower bound is achieved, then $L(\mathbf{x}; \theta)$ belongs to the exponential family (§3.32) and is of the form:

$$L(\mathbf{x}; \theta) = c(\theta) \exp[b(\theta) T(\mathbf{x})] h(\mathbf{x}).$$

Conversely, if $L(\mathbf{x}; \theta)$ in (i) belongs to the exponential family as given above, $b(\theta)$ has a continuous nonvanishing derivative on Ω, then $\sigma^2(T) = (g'(\theta))^2/I(\theta)$, and if the range of $b(\theta)$ contains an open interval then $T(\mathbf{X})$ is a uniformly minimum variance unbiased (UMVU) estimate of $g(\theta)$.

For example, if each X_i has a $N(\mu, 1)$ distribution, then with $\theta = \mu$, $E_\theta[((\partial/\partial\theta) \ln f(x; \theta))^2] = 1$, and $1/I(\theta) = 1/n$, $\sigma^2(\bar{X}) = 1/n$. Hence $\bar{X}$ is a UMVU estimator of μ.

As another example, suppose each X_i has a binomial distribution with parameters "1" and "p." Then

$$f(x; p) = (p)^x(1 - p)^{1-x}, \qquad \frac{\partial}{\partial p} \ln f(x; p) = \frac{x}{p} - \frac{1 - x}{1 - p},$$

and $1/I(\theta) = p(1 - p)/n$. Also $\sigma^2(\bar{X}) = p(1 - p)/n$, therefore $\bar{X}$ is a UMVU estimator of p.

[Cf., Roussas (1973), Bickel and Doksum (1977), Fourgeaud and Fuchs (1967).]

§1.18. Sufficient Statistics

Let $X_1, \ldots, X_n$ be independent identically distributed random variables each with a distribution F_θ depending on a parameter θ. A statistic $T(X_1, \ldots, X_n) \equiv T(\mathbf{X})$ is called a *sufficient statistic* for the parameter θ, if and only if, the conditional distribution of $\mathbf{X} = (X_1, \ldots, X_n)'$ given that $T(\mathbf{X}) = t$ is independent of θ. Hence the statistic T makes use of all the information in the sample regarding the parameter θ, and once a value of T is given, the sample does not contain any additional information about the parameter θ.

For example, for the exponential family (see §3.32) with probability density or probability mass function:

$$f(\mathbf{x}; \theta) = (a(\theta))^n \exp\left[b(\theta) \sum_{i=1}^{n} t(x_i)\right] \prod_{i=1}^{n} h(x_i),$$

where $a(\theta) > 0$, and the set S of positivity of $f(\mathbf{x}; \theta)$ is independent of θ, $T(\mathbf{X}) = \sum_{i=1}^{n} t(X_i)$ is a sufficient statistic for θ. For the discrete case, the probability mass function of $T(\mathbf{X})$ is given by

$$(a(\theta))^n \exp[b(\theta)t] \sum_{S_0} h(x_1) \ldots h(x_n),$$

where the sum is over all $x_1, \ldots, x_n$ in the set $S_0 \subseteq S$ such that $t(x_1) + \cdots + t(x_n) = t$. For the continuous case, it is sufficient to assume that $y = t(x)$ is one-to-one, $\partial x_i/\partial y_1$ exists and is continuous for all $i = 1, \ldots, n$, where $y_1 = \sum_{i=1}^{n} t(x_i)$. The probability density of $T(\mathbf{X})$ is then,

$$(a(\theta))^n \exp[b(\theta)t] \int_{S_0} \left(\prod_{i=1}^{n} h(x_i)\right) \left|\frac{\partial x_1}{\partial y_1}\right| dx_2 \ldots dx_n,$$

where x_1 is to be expressed in terms of $t, x_2, \ldots, x_n$, and S_0 is the image of the set of positivity of $h(x_1), \ldots, h(x_n)$ under T. [For the k-parameter exponential family (see §3.32)

$$f(\mathbf{x}; \boldsymbol{\theta}) = a(\boldsymbol{\theta}) \exp\left[\sum_{i=1}^{k} c_i(\boldsymbol{\theta}) t_i(\mathbf{x})\right] h(\mathbf{x}), \qquad \boldsymbol{\theta} = (\theta_1, \ldots, \theta_k)', \quad \mathbf{x} = (x_1, \ldots, x_n)',$$

the vector $(t_1(\mathbf{x}), \ldots, t_k(\mathbf{x}))'$ is a sufficient statistic for $\boldsymbol{\theta}$.]

[Cf., Fourgeaud and Fuchs (1967), Roussas (1973), Bickel and Doksum (1977).]

§1.19. Fisher–Neyman Factorization Theorem for Sufficient Statistics

Let $X_1, \ldots, X_n$ be independent identically distributed random variables with joint probability density or probability mass function $f(\mathbf{x}; \theta)$ depending on a parameter θ. A statistic $T(\mathbf{X})$ is sufficient for the parameter θ if and only if $f(\mathbf{x}; \theta)$ may be factored out as:

$$f(\mathbf{x}; \theta) = g(T(\mathbf{x}); \theta) h(\mathbf{x}),$$

where $h(\mathbf{x})$ is independent of θ, g depends on $\mathbf{x}$ only through T, and $h(\mathbf{x})$ may be so chosen that $g(t; \theta)$ denotes the probability density or probability mass function of T.

EXAMPLES

(i) Suppose each X_i has an exponential distribution with density $\alpha e^{-\alpha x}$. Then

$$\begin{aligned} f(\mathbf{x}; \alpha) &= (\alpha)^n \exp[-\alpha(x_1 + \cdots + x_n)], \qquad 0 < x_i < \infty, \quad i = 1, \ldots, n \\ &= \left\{(\alpha)^n \frac{(t)^{n-1}}{(n-1)!} e^{-\alpha t}\right\} \cdot \frac{(n-1)!}{(x_1 + \cdots + x_n)^{n-1}}, \end{aligned}$$

where $t = x_1 + \cdots + x_n$, and the expression in the curly brackets is its associated density (see §§3.20 and 4.18). Hence $\sum_{i=1}^{n} X_i$ is a sufficient statistic for $1/\alpha$.

(ii) Suppose each X_i has a normal distribution with mean μ and variance 1. Then

$$\begin{aligned} f(\mathbf{x}; \mu) &= (2\pi)^{-n/2} \exp\left[-\sum_{i=1}^{n} (x_i - \mu)^2/2\right] \\ &= \left\{\sqrt{\frac{n}{2\pi}} \exp[-n(\bar{x} - \mu)^2/2]\right\} \cdot \frac{(2\pi)^{-(n-1)/2}}{\sqrt{n}} \exp\left[-\sum_{i=1}^{n} (x_i - \bar{x})^2/2\right], \end{aligned}$$

where $\bar{x} = \sum_{i=1}^{n} x_i/n$, and the expression in the curly brackets is its associated density. Hence $\sum_{i=1}^{n} X_i/n$ is a sufficient statistic for μ.

(iii) Suppose each X_i has a normal distribution with mean 0 and variance σ^2. Then

$$f(\mathbf{x}; \sigma^2) = (2\pi\sigma^2)^{-n/2} \exp\left(-\sum_{i=1}^{n} x_i^2/2\sigma^2\right)$$

$$= \left\{\frac{1}{2^{n/2}\Gamma(n/2)}\left(\frac{t}{\sigma^2}\right)^{(n-2)/2} \frac{e^{-t/2\sigma^2}}{\sigma^2}\right\} \times \left(\frac{1}{\pi}\right)^{n/2} \frac{\Gamma(n/2)}{\left(\sum_{i=1}^{n} x_i^2\right)^{(n-2)/2}},$$

where $t = \sum_{i=1}^{n} x_i^2$, and the expression in the curly brackets is its associated density (see §§3.35 and 4.25), and hence $\sum_{i=1}^{n} X_i^2$ is a sufficient statistic for σ^2.

(iv) Suppose each X_i has a normal distribution with mean μ and variance σ^2. Let $\boldsymbol{\theta} = (\mu, \sigma^2)'$. Then

$$f(\mathbf{x}; \sigma^2) = (2\pi\sigma^2)^{-n/2} \exp\left[-\sum_{i=1}^{n} (x_i - \mu)^2/2\sigma^2\right]$$

$$= \left\{\sqrt{\frac{n}{2\pi\sigma^2}} \exp[-n(t_1 - \mu)^2/2\sigma^2] \cdot \frac{1}{2^{(n-1)/2}} \frac{1}{\Gamma\left(\frac{n-1}{2}\right)}\right.$$

$$\left. \times \left(\frac{t_2}{\sigma^2}\right)^{(n-3)/2} \frac{e^{-t_2/2\sigma^2}}{\sigma^2}\right\} \times \left(\frac{1}{\pi}\right)^{(n-1)/2} \frac{\Gamma\left(\frac{n-1}{2}\right)}{(t_2)^{(n-3)/2}},$$

where $t_1 = \bar{x}$, $t_2 = \sum_{i=1}^{n} (x_i - \bar{x})^2$, and the expression in the curly brackets is the density of $(\bar{X}, \sum_{i=1}^{n} (X_i - \bar{X})^2)'$, and hence the latter random vector is a sufficient statistic for $\boldsymbol{\theta} = (\mu, \sigma^2)'$.

(v) Suppose each X_i has a Poisson distribution with mean λ. Then

$$f(\mathbf{x}; \lambda) = \frac{e^{-\lambda n} \exp[(\ln \lambda)(x_1 + \cdots + x_n)]}{x_1! \ldots x_n!}$$

$$= \left\{\frac{e^{-n\lambda}(n\lambda)^t}{t!}\right\} \frac{(x_1 + \cdots + x_n)!}{x_1! \ldots x_n!} n^{-(x_1 + \cdots + x_n)},$$

where $t = x_1 + \cdots + x_n$, and the expression in the curly brackets is its associated density (see §4.13), and hence $\sum_{i=1}^{n} X_i$ is a sufficient statistic for λ.

[Cf., Hogg and Craig (1978), Fourgeaud and Fuchs (1967).]

§1.20. Rao–Blackwell Theorem

Let $X_1, \ldots, X_n$ be independent identically distributed random variables each with a distribution F_θ depending on a parameter $\theta \in \Omega$. Suppose $T(X_1, \ldots, X_n) \equiv T(\mathbf{X})$ is a sufficient statistic for θ, and $S(X_1, \ldots, X_n) \equiv S(\mathbf{X})$, is any (other) statistic which is an unbiased estimator of θ. Define the function $G(t) = E_\theta[S(\mathbf{X})|T(\mathbf{X}) = t]$. [Note that since T is sufficient, $G(t)$ is independent of θ, and hence $G(T)$ is a statistic.] Then $E_\theta[G(T)] = \theta$, that is $G(T)$ is an unbiased estimator of θ, and $E_\theta[(G(T) - \theta)^2] \leqslant E_\theta[(S(\mathbf{X}) - \theta)^2]$ for all $\theta \in \Omega$. If $E_\theta[(S(\mathbf{X}) - \theta)^2] < \infty$, then we have a strict inequality $E_\theta[(G(T) - \theta)^2] < E_\theta[(S(\mathbf{X}) - \theta)^2]$ unless $G(T) = S(\mathbf{X})$ (with probability one). [The theorem is also true if $S(\mathbf{X})$ is an unbiased estimator of a function $g(\theta)$ of θ, then we have $E_\theta[G(T)] = g(\theta)$, and

$$E_\theta[(G(T) - g(\theta))^2] \leqslant E_\theta[(S(\mathbf{X}) - g(\theta))^2].]$$

[Cf., Bickel and Doksum (1977), Roussas (1973).]

§1.21. Completeness of Statistics and Their Families of Distributions

Let $X_1, \ldots, X_n$ be independent identically distributed random variables each with a distribution F_θ depending on a parameter $\theta \in \Omega$, where Ω is the parameter space. A statistic $T(X_1, \ldots, X_n) \equiv T(\mathbf{X})$ is said to be *complete* if for every function $g(T)$ such that $E_\theta[g(T)] = 0$ for all $\theta \in \Omega$ implies that $P_\theta[g(T) = 0] = 1$ for all $\theta \in \Omega$ (that is, $g(T) = 0$ with probability (P_θ)—one for all $\theta \in \Omega$). The family of distributions of T as θ varies in Ω is also called a *complete family* of distributions if T is complete.

§1.22. Theorem on Completeness of Statistics with Sampling from the Exponential Family

Let $X_1, \ldots, X_n$ be independent identically distributed random variables with joint probability density or mass function $f(\mathbf{x}; \theta)$ belonging to the exponential family (see §3.32):

$$f(\mathbf{x}; \theta) = (a(\theta))^n \exp\left[b(\theta) \sum_{i=1}^{n} t(x_i)\right] \prod_{i=1}^{n} h(x_i),$$

where $\theta \in \Omega \subseteq R^1$. Then the statistic $T(\mathbf{X}) = \sum_{i=1}^n t(X_i)$ is complete provided the range of $b(\theta)$ contains an open interval (b_1, b_2), $b_1 < b_2$. For the k-parameter family:

$$f(\mathbf{x}; \boldsymbol{\theta}) = a(\boldsymbol{\theta}) \exp\left[\sum_{i=1}^{k} c_i(\boldsymbol{\theta}) t_i(\mathbf{x})\right] h(\mathbf{x}),$$

where $\theta \in \Omega \subseteq R^k$, $\mathbf{x} = (x_1, \ldots, x_n)'$. Then the vector statistic $(t_1(\mathbf{X}), \ldots, t_k(\mathbf{X}))'$ is complete if the range of the vector $(c_1(\boldsymbol{\theta}), \ldots, c_k(\boldsymbol{\theta}))'$ contains an open k-rectangle, that is a set of the form: $\{(c_1, \ldots, c_k): A_i < c_i < B_i, i = 1, \ldots, k\}$.

[Cf., Lehmann (1959).]

EXAMPLES

(i) Suppose each X_i has a Poisson distribution with mean $\lambda \in \Omega = \{\lambda: 0 < \lambda < \infty\}$. Then

$$f(\mathbf{x}; \lambda) = \frac{e^{-n\lambda} . e^{(x_1 + \cdots + x_n) \ln \lambda}}{x_1! \ldots x_n!},$$

$-\infty < \ln \lambda < \infty$ which contains an open interval, and hence $\sum_{i=1}^n X_i$ is a complete statistic.

(ii) Suppose each X_i has a $N(\mu, \sigma^2)$ distribution with $\boldsymbol{\theta} = (\theta_1, \theta_2)' \in \Omega = \{\boldsymbol{\theta}: -\infty < \theta_1 < \infty, 0 < \theta_2 < \infty\}$, where $\theta_1 = \mu, \theta_2 = \sigma^2$. Then

$$f(\mathbf{x}; \boldsymbol{\theta}) = (2\pi\sigma^2)^{-n/2} e^{-n\mu^2/2\sigma^2} \exp\left[\frac{\mu}{\sigma^2} t_1 - \frac{1}{2\sigma^2} t_2\right],$$

where $t_1 = \sum_{i=1}^n x_i$, $t_2 = \sum_{i=1}^n x_i^2$, $c_1(\boldsymbol{\theta}) = \mu/\sigma^2$, $c_2(\boldsymbol{\theta}) = -1/2\sigma^2$. The vector $(c_1(\boldsymbol{\theta}), c_2(\boldsymbol{\theta}))'$ ranges over the lower half-plane as μ and σ^2 vary in Ω. Hence $(\sum_{i=1}^n X_i, \sum_{i=1}^n X_i^2)'$ is complete. If Ω was, for example, in the form $\Omega' = \{\boldsymbol{\theta}: \mu = \sigma^2, 0 < \sigma^2 < \infty\}$, then $c_1(\boldsymbol{\theta}) = 1$, $c_2(\boldsymbol{\theta}) = -1/2\sigma^2$, and $\{(c_1, c_2): 0 < \sigma^2 < \infty\}$ which contains no open two-dimensional rectangle and hence one cannot use the above theorem to make a statement about the statistic $\mathbf{T} = (\sum_{i=1}^n X_i, \sum_{i=1}^n X_i^2)'$, concerning completeness, in reference to Ω'. As a matter of fact, let $g(\mathbf{T}) = (T_2/(n-1)) - (T_1^2/n(n-1)) - (T_1/n)$, where $T_1 = \sum_{i=1}^n X_i$, $T_2 = \sum_{i=1} X_i^2$. We note we may rewrite

$$g(\mathbf{T}) = \left[\sum_{i=1}^n (X_i - \bar{X})^2/(n-1)\right] - \bar{X}, \quad \text{and} \quad E_{\boldsymbol{\theta}}[g(\mathbf{T})] = \sigma^2 - \mu = 0$$

for all $\boldsymbol{\theta} \in \Omega'$. However, $g(\mathbf{T})$ is not in general equal to zero. Accordingly, $\mathbf{T}$, in reference to Ω', is not complete.

(iii) Suppose each X_i has a normal distribution with mean $\mu \in \Omega = \{\mu: -\infty < \mu < \infty\}$, and variance 1, then

$$f(\mathbf{x}; \mu) = (2\pi)^{-n/2} \exp(-n\mu^2/2) \exp\left(-\sum_{i=1}^n x_i^2/2\right) \exp\left(\mu \sum_{i=1}^n x_i\right)$$

and hence $\sum_{i=1}^n X_i$ is complete.

(iv) Suppose each X_i has a normal distribution with mean 0 and variance $\sigma^2 \in \Omega = \{\sigma^2: 0 < \sigma^2 < \infty\}$, then

$$f(\mathbf{x}; \mu) = (2\pi\sigma^2)^{-n/2} \exp\left(-\frac{1}{2\sigma^2} \sum_{i=1}^n x_i^2\right),$$

$-\infty < -1/2\sigma^2 < 0$, and hence $\sum_{i=1}^n X_i^2$ is complete.

§1.23. Lehmann–Scheffé Uniqueness Theorem

Let $X_1, \ldots, X_n$ be independent identically distributed random variables each with a distribution depending on a parameter $\theta \in \Omega$. Suppose $T(X_1, \ldots, X_n) \equiv T(\mathbf{X})$ is a complete sufficient statistic and $S(X_1, \ldots, X_n) \equiv S(\mathbf{X})$ is an unbiased estimator of $g(\theta)$. Then $G(T) = E_\theta[S(\mathbf{X}) | T(\mathbf{X})]$ is a uniformly minimum variance unbiased (UMVU) estimator of $g(\theta)$. Also if $E_\theta[(G(T) - g(\theta))^2] < \infty$, then $G(T)$ is the unique UMVU estimator of $g(\theta)$. In particular, we note that if $S(\mathbf{X})$ is of the form $H(T)$, that is, it is a function of T, then $E_\theta[H(T)|T] = H(T)$ and $H(T)$ is a UMVU estimator of $g(\theta)$.

Examples

(i) Suppose each X_i has a normal distribution with mean $\mu \in \Omega = \{\mu: -\infty < \mu < \infty\}$ and variance 1, then we know (see §§1.19 and 1.22) that $\bar{X} = \sum_{i=1}^n X_i/n$ is sufficient and complete. It is also an unbiased estimator of μ and hence it is a UMVU estimator of μ.

(ii) Suppose each X_i has a normal distribution with mean 0 and variance $\sigma^2 \in \Omega = \{\sigma^2: 0 < \sigma^2 < \infty\}$, then we know (see §§1.19 and 1.22) that $\sum_{i=1}^n X_i^2$ is sufficient and complete. Also $\sum_{i=1}^n X_i^2/n$ is an unbiased estimator of σ^2, and hence it is a UMVU estimator of σ^2.

[Cf., Bickel and Doksum (1977), Roussas (1973).]

§1.24. Efficiency, Relative Efficiency, and Asymptotic Efficiency of Estimators

(i) Let $X_1, \ldots, X_n$ be independent identically distributed random variables each with a distribution F_θ depending on a parameter. Suppose that $T_n(X_1, \ldots, X_n) \equiv T_n(\mathbf{X})$ is an unbiased estimator of $g(\theta)$ such that the Cramér–Rao–Frêchet inequality (§1.17): $\sigma^2(T_n) \geqslant (g'(\theta))^2/I(\theta)$ holds. The *efficiency* of the statistic T in estimating $g(\theta)$ is defined by $e(T_n) = (g'(\theta))^2/(I(\theta)\sigma^2(T_n))$. Clearly, $e(T_n)$ is a number between 0 and 1. The asymptotic efficiency of T_n is defined by $\lim_{n\to\infty} e(T_n)$. If the latter is equal to one then T_n is called asymptotically efficient.

(ii) Suppose $\{T_{1n}\}$ and $\{T_{2n}\}$ are two sequences of asymptotically unbiased estimators (§1.9.) of the parameter $g(\theta)$, and $\{\sigma_{1n}^2\}$, $\{\sigma_{2n}^2\}$ denote the corresponding asymptotic variances. Suppose that $n\sigma_{1n}^2 \to \sigma_1^2$, $n\sigma_{2n}^2 \to \sigma_2^2$, for $n \to \infty$. Then the *asymptotic efficiency* of T_{1n} to T_{2n} is defined by $e_\infty(T_1, T_2) = \sigma_2^2/\sigma_1^2$.

(iii) If T_{1n} and T_{2n} are two unbiased estimators of a parameter $g(\theta)$, then the *relative efficiency* of T_{1n} to T_{2n} is defined by $e_n(T_1, T_2) = \sigma^2(T_{2n})/\sigma^2(T_{1n})$.

§1.25. Estimation by the Method of Moments

(i) Let $X_1, \ldots, X_n$ be independent identically distributed random variables each with a distribution F_θ depending on a parameter θ. Suppose θ may be written as a function of the first r moments $\alpha_1, \ldots, \alpha_r$ of the population (see §1.3.): $\theta = \theta(\alpha_1, \ldots, \alpha_r)$, $\alpha_i = E[X_1^i]$. Let $a_1, \ldots, a_r$ denote the corresponding sample moments (§1.3): $a_i = \sum_{j=1}^{n} X_j^i/n$. Then the method of moments of estimation of θ is given by $\hat{\theta} = \theta(a_1, \ldots, a_r)$, by replacing α_i by a_i for $i = 1, \ldots, r$, in θ. It does not always provide a unique estimator.

(ii) *For Grouped Data* (see §2.23). We group the data $X_1, \ldots, X_n$ into mutually exclusive (intervals) classes $c_1, \ldots, c_k$. Let N_i denote the number of the X_i falling in c_i. A class c_i is of the form $[b_{i-1}, b_i)$, $i = 1, \ldots, k$. Let $Y_i = (b_i + b_{i-1})/2$, and define

$$d_t = \sum_{i=1}^{k} N_i Y_i^t/n, \quad \text{where} \quad \sum_{i=1}^{k} N_i = n.$$

Then the method of moments of estimation from grouped data may be defined by $\hat{\theta} = \theta(d_1, \ldots, d_r)$. Variations of this definition are also possible.

§1.26. Confidence Intervals

Let $X_1, \ldots, X_n$ be independent identically distributed random variables each having a distribution F_θ depending on a parameter $\theta \in \Omega$ with a given parameter space Ω. Let $\underline{L} = \underline{L}(X_1, \ldots, X_n)$ and $\bar{L} = \bar{L}(X_1, \ldots, X_n)$ be two statistics of the X_i. Then the interval $[\underline{L}, \bar{L}]$, also called a random interval, is a *confidence interval* for θ with *confidence coefficient* $1 - \alpha$ $(0 < \alpha < 1)$, if $P_\theta[\underline{L}(X_1, \ldots, X_n) \leqslant \theta \leqslant \bar{L}(X_1, \ldots, X_n)] \geqslant 1 - \alpha$ for all $\theta \in \Omega$. That is, the interval $[\underline{L}, \bar{L}]$ covers the parameter θ, with probability at least equal to $1 - \alpha$, for *all* $\theta \in \Omega$. $\underline{L}$, $\bar{L}$ are, respectively, called the lower and upper confidence limits. (In some applications, one may replace either $\underline{L}$ or $\bar{L}$ by $-\infty$ or $+\infty$, respectively.)

EXAMPLES

(i) Suppose that each X_i has a normal distribution with mean μ and variance σ^2. If σ^2 is known then a $(1 - \alpha)\%$ confidence interval for μ is given by $[\underline{L}, \bar{L}]$, where

$$\underline{L} = \bar{X} - Z_{\alpha/2}\frac{\sigma}{\sqrt{n}}, \qquad \bar{L} = \bar{X} + Z_{\alpha/2}\frac{\sigma}{\sqrt{n}},$$

where $Z_{1-\alpha/2}$ denotes the $(\alpha/2)$th quantile of the standard normal distribution.

If σ^2 is unknown then

$$\underline{L} = \bar{X} - t_{\alpha/2}(n-1)\frac{S}{\sqrt{n}}, \qquad \bar{L} = \bar{X} + t_{\alpha/2}(n-1)\frac{S}{\sqrt{n}},$$

where

$$S^2 = \sum_{i=1}^{n} (X_i - \bar{X})^2/(n-1),$$

and $t_{\alpha/2}(n-1)$ is the $(1-(\alpha/2))$th quantile of the Student distribution of $n-1$ degrees of freedom. A $(1-\alpha)\%$ confidence interval for σ^2 is given by $[\underline{L}, \bar{L}]$, where

$$\underline{L} = (n-1)S^2/\chi^2_{\alpha/2}(n-1), \qquad \bar{L} = (n-1)S^2/\chi^2_{1-\alpha/2}(n-1),$$

where $\chi^2_{1-\alpha/2}(n-1)$ is the $(\alpha/2)$th quantile of the chi-square distribution of $(n-1)$ degrees of freedom. If the normality condition of the X_i is in doubt then an approximate, though robust, confidence interval for σ^2 may be directly developed by the method of jackknife (see §§1.41 and 2.33).

(ii) For the parameter "p" of a binomial distribution we have (see also §4.4):

$$\underline{L} = \frac{x}{x + (n-x+1)F_{\alpha/2}(2(n-x+1), 2x)},$$

$$\bar{L} = \frac{(x+1)F_{\alpha/2}(2(x+1), 2(n-x))}{(n-x) + (x+1)F_{\alpha/2}(2(x+1), 2(n-x))},$$

where x is the number of "successes" obtained in n trials, and $F_{1-\alpha/2}(\nu_1, \nu_2)$ is the $(\alpha/2)$th quantile of the F-distribution of ν_1 and ν_2 degrees of freedom. Here each X_i has a binomial distribution with parameters "1" and "p."

(iii) For the parameter λ of a Poisson distribution (see §4.14) we have

$$\bar{L} = \tfrac{1}{2}\chi^2_{\alpha/2}(2x), \qquad \underline{L} = \tfrac{1}{2}\chi^2_{1-\alpha/2}(2(x+1)),$$

where x denotes the value obtained for a random variable distributed according to a Poisson distribution.

(iv) Let $\mathbf{X}_1, \ldots, \mathbf{X}_n$ be independent identically distributed random k-vectors each with a $N(\boldsymbol{\mu}, \boldsymbol{\Sigma})$ distribution with *known* covariance matrix $\boldsymbol{\Sigma}$. Then a $(1-\alpha)100\%$ confidence region for $\boldsymbol{\mu}$ is given by the set of *all* k-vectors $\mathbf{m}$ satisfying:

$$n(\bar{\mathbf{X}} - \mathbf{m})'\boldsymbol{\Sigma}^{-1}(\bar{\mathbf{X}} - \mathbf{m}) \leqslant \chi^2_{\alpha}(k),$$

where $\bar{\mathbf{X}} = \sum_{i=1}^{n} \mathbf{X}_i/n$, and $\chi^2_{\alpha}(k)$ denotes the $(1-\alpha)$th quantile of the chi-distribution of k degrees of freedom.

(v) Let $\mathbf{X}_1, \ldots, \mathbf{X}_n$ be as in (iv) except suppose that $\boldsymbol{\Sigma}$ is unknown. Let $\mathbf{S} = \sum_{i=1}^{n} (\mathbf{X}_i - \bar{\mathbf{X}})(\mathbf{X}_i - \bar{\mathbf{X}})'$. Then a $(1-\alpha)100\%$ confidence region for $\boldsymbol{\mu}$ is given by the set of *all* k-vectors $\mathbf{m}$ satisfying:

$$\frac{(n-k)}{k(n-1)} n(\bar{\mathbf{X}} - \mathbf{m})' \mathbf{S}^{-1} (\bar{\mathbf{X}} - \mathbf{m}) \leqslant F_\alpha(k, n-k),$$

where $F_\alpha(k, n-k)$ is the $(1-\alpha)$th quantile of Fisher's F-distribution of k and $(n-k)$ degrees of freedom $(k < n)$. A $(1-\alpha)100\%$ confidence region (interval) may be also given for any linear function $\mathbf{a}'\boldsymbol{\mu}$ where $\mathbf{a}$ is any k-vector. The latter is given by

$$\mathbf{a}'\bar{\mathbf{X}} - T_\alpha \frac{(\mathbf{a}'\mathbf{S}\mathbf{a})^{1/2}}{n^{1/2}} \leqslant \mathbf{a}'\boldsymbol{\mu} \leqslant \mathbf{a}'\bar{\mathbf{X}} + T_\alpha \frac{(\mathbf{a}'\mathbf{S}\mathbf{a})^{1/2}}{n^{1/2}},$$

where

$$T_\alpha = \sqrt{\frac{k(n-1)}{(n-k)} F_\alpha(k, n-k)}.$$

§1.27. Tolerance Intervals

Let $X_1, \ldots, X_n$ be independent identically distributed random variables each with a distribution $F(x)$.

(i) Let $\underline{L} = \underline{L}(X_1, \ldots, X_n)$, $\bar{L} = \bar{L}(X_1, \ldots, X_n)$ be two statistics of the X_i. The interval $(\underline{L}, \bar{L}]$ is called a $(1-\alpha)100\%$ $(0 < \alpha < 1)$ tolerance interval of $p100\%$ $(0 < p < 1)$ of F if $P[F(\bar{L}) - F(\underline{L}) \geqslant p] \geqslant 1 - \alpha$.

(ii) In particular, if the X_i are continuous random variables, then $(X_{(a)}, X_{(b)}]$, $a < b$, where $X_{(1)} \leqslant \cdots \leqslant X_{(n)}$ are the order statistics of the X_i, is a $(1-\alpha)100\%$ tolerance interval of $p100\%$ of F with

$$1 - \alpha = \frac{n!}{(b-a-1)!(n-b+a)!} \int_p^1 x^{b-a-1}(1-x)^{n-b+a}\, dx.$$

[Cf., Roussas (1973).]

§1.28. Simple and Composite Hypotheses, Type-I and Type-II Errors, Level of Significance or Size, Power of a Test and Consistency

Let X be a random variable with distribution F_θ depending on some parameter θ which may have a value in some set Ω: $\theta \in \Omega$. The set Ω is called the parameter space. A hypothesis or an assertion about the parameter θ is called a statistical hypothesis about θ. We consider the test of hypothesis that θ belongs to some subset ω of Ω: H_0: $\theta \in \omega \not\subseteq \Omega$ against a hypothesis that θ does not belong to ω: H_A: $\theta \in \Omega - \omega$. The hypotheses H_0 and H_A are, respectively, called the null and the alternative hypotheses. In testing the hypothesis H_0: $\theta \in \omega$ against H_A: $\theta \in \Omega - \omega$, the probability P_θ[rejecting H_0], for $\theta \in \omega$, is called a Type-I error, and the probability P_θ[accepting H_0], for $\theta \in \Omega - \omega$, is called a Type-II error and is denoted by $\beta(\theta)$. The probability P_θ[rejecting H_0] for $\theta \in \Omega - \omega$, is called the power of the test at θ. Finally, $\sup_{\theta \in \omega} P_\theta$[rejecting H_0] is called the level of significance or size of the test and is denoted by α. Suppose n denotes the size of the sample in question used in carrying out the test. If for all $\theta \in \Omega - \omega$, $\lim_{n \to \infty} P_\theta$[rejecting H_0] $= 1$, then the test is said to be consistent. [A test of hypothesis is not necessarily restricted to making an assertion about a parameter θ and may involve, for example, in making an assertion about the distribution F itself such as H_0: $F(x) \equiv G(x)$ against H_A: $F(x) \not\equiv G(x)$, where $G(x)$ some given distribution function.] If ω consists only of one element $\{\theta_0\}$, then H_0 is called a simple hypothesis otherwise it is called a composite hypothesis. A similar definition holds for H_A.

§1.29. Randomized and Nonrandomized Test Functions

Let $X_1, \ldots, X_n$ be independent identically distributed random variables. A real function $\phi(x_1, \ldots, x_n) \equiv \phi(\mathbf{x})$ for testing a certain hypothesis H_0 against an alternative hypothesis H_A is called a randomized test if $\phi(\mathbf{x}) = \delta$ where $\mathbf{x}$ is the observed value of $\mathbf{X}$, with $0 \leqslant \delta \leqslant 1$, and if a coin with probability equal to δ of falling head is tossed and a head (tail) is obtained in a trial, then H_0 is rejected (accepted). If δ is either 0 or 1, then $\phi(\mathbf{x})$ is called a nonrandomized test function for testing the hypothesis H_0 against H_A. If we consider a test of hypothesis about a parameter $\theta \in \Omega$: H_0: $\theta \in \omega \nsubseteq \Omega$ against H_A: $\theta \in \Omega - \omega$, then we may simply write P_θ[rejecting H_0] $= E_\theta[\phi(\mathbf{X})]$.

§1.30. Uniformly Most Powerful (UMP), Most Powerful (MP), Unbiased and Uniformly Most Powerful Unbiased (UMPU) Tests

A size α-test which maximizes the power among all size α-tests is said to be UMP. For example, consider a test of hypothesis about a parameter $\theta \in \Omega$: H_0: $\theta \in \omega$ against H_A: $\theta \in \Omega - \omega$. Then a randomized test function ϕ is a UMP size α-test if $\sup_{\theta \in \omega} E_\theta[\phi] = \alpha$, and for any randomized test function $\tilde{\phi}$ size α-test ($\sup_{\theta \in \omega} E_\theta[\tilde{\phi}] = \alpha$), $E_\theta[\phi] \geqslant E_\theta[\tilde{\phi}]$ for all $\theta \in \Omega - \omega$. If the set $\Omega - \omega$ consists only of one point, then a UMP test is simply termed most powerful (MP). A size α-test ϕ is said to be unbiased if $E_\theta[\phi] \geqslant \alpha$ for $\theta \in \Omega - \omega$. A size α-test ϕ is said to be uniformly most powerful unbiased (UMPU) test if it is UMP within the class of all unbiased size α-tests.

§1.31. Neyman–Pearson Fundamental Lemma

(i) (Neyman–Pearson). Let $X_1, \ldots, X_n$ be independent identically distributed random variables each with a probability density or probability mass function $f(x; \theta)$ depending on a parameter θ (may be a vector). Consider the test of hypothesis $H_0: \theta = \theta_0$ against $H_A: \theta = \theta_1$. Then for $0 < \alpha < 1$, the most powerful (MP) size α-test is given by

$$\phi(\mathbf{x}) = \begin{cases} 1 & \text{if } L(\theta_1; \mathbf{x}) > kL(\theta_0; \mathbf{x}), \\ \delta & \text{if } L(\theta_1; \mathbf{x}) = kL(\theta_0; \mathbf{x}), \\ 0 & \text{if } L(\theta_1; \mathbf{x}) < kL(\theta_0; \mathbf{x}), \end{cases}$$

where $L(\theta; \mathbf{x}) = \prod_{i=1}^n f(x_i; \theta)$ is the likelihood function, and $0 \leqslant \delta \leqslant 1$, $k > 0$ are determined so that $E_{\theta_0}[\phi(\mathbf{X})] = \alpha$. Also the test is unbiased: $E_{\theta_1}[\phi(\mathbf{X})] \geqslant \alpha$, that is the probability of rejecting H_0, if H_0 is false, is not any smaller than the probability of rejecting H_0, if H_0 is true.

(ii) Let $g(x; \theta_0, \theta_1) = \ln[f(x; \theta_1)/f(x; \theta_0)]$. Suppose $E_\theta[|g(X; \theta_0, \theta_1)|] < \infty$ for $\theta = \theta_0$ and for $\theta = \theta_1$. If $E_{\theta_0}[g(X; \theta_0, \theta_1)] < E_{\theta_1}[g(X; \theta_0, \theta_1)]$, then the test is consistent, that is, $\lim_{n\to\infty} P_{\theta_1}[\phi(\mathbf{X})] = 1$.

[Cf., Lehmann (1959), Barra (1981), Zacks (1971).] See also §§1.32, 1.33, 1.34, 1.35, 1.36, 1.38, 1.39.

§1.32. Monotone Likelihood Ratio Property of Family of Distributions and Related Theorems for UMP and UMPU Tests for Composite Hypotheses

(i) Let $X_1, \ldots, X_n$ be independent identically distributed random variables each with a probability density or probability mass $f(x; \theta)$, where $\theta \in \Omega \subseteq R^1$. Consider the family $\{L(\theta; \mathbf{x}); \theta \in \Omega\}$, where $L(\theta; \mathbf{x})$ is the likelihood function $\prod_{i=1}^n f(x_i; \theta)$, $\mathbf{x} = (x_1, \ldots, x_n)'$. Suppose the set $S = \{\mathbf{x}: L(\theta; \mathbf{x}) > 0\}$ of positivity of $L(\theta; \mathbf{x})$ is independent of θ for all $\theta \in \Omega$. If there exists a real function $Q(\mathbf{x})$, independent of θ, with $\mathbf{x} \in S$, such that for all $\theta_1 < \theta_2$ with $\theta_1, \theta_2 \in \Omega$, then the family $\{L(\theta; \mathbf{x}); \theta \in \Omega\}$ is said to have a monotone likelihood ratio property in Q if $L(\theta_1; \mathbf{x}) \neq L(\theta_2; \mathbf{x})$ and $L(\theta_2; \mathbf{x})/L(\theta_1; \mathbf{x})$ is a strictly increasing function of Q, for almost all $x \in S$. $L(\theta_2; \mathbf{x})/L(\theta_1; \mathbf{x})$ is called the likelihood ratio.

(ii) For the exponential family (§3.32(I)):

$$L(\theta; \mathbf{x}) = (a(\theta))^n \exp[b(\theta) T(\mathbf{x})] \prod_{i=1}^n h(x_i),$$

where $T(\mathbf{x}) = \sum_{i=1}^n t(x_i)$, if $b(\theta_2) > b(\theta_1)$, for all $\theta_2 > \theta_1$, $\theta_1, \theta_2 \in \Omega \subseteq R^1$, then the exponential family has a monotone likelihood ratio property in $Q(\mathbf{x}) =$

$T(\mathbf{x})$, and if $b(\theta_2) < b(\theta_1)$, for all $\theta_2 > \theta_1, \theta_1, \theta_2 \in \Omega \subseteq R^1$, then the exponential family has a monotone likelihood ratio propery in $Q(\mathbf{x}) = -T(\mathbf{x})$.

(iii) Let $X_1, \ldots, X_n$ be independent identically distributed random variables each with a probability density or mass function $f(x; \theta)$. Suppose the family $\{L(\theta; \mathbf{x}); \theta \in \Omega\}$ has a monotone likelihood ratio property in $Q(\mathbf{x})$. Consider the test of hypothesis $H_0: \theta \in \omega \equiv \{\theta \in \Omega, \theta \leqslant \theta_0\}$ against $H_A: \theta \in \Omega - \omega$, for some fixed $\theta_0 \in \Omega$. Then for $0 < \alpha < 1$, there exists a uniformly most powerful (UMP) size α-test within the class of all size $\leqslant \alpha$-tests. It is given by

$$\phi(\mathbf{x}) = \begin{cases} 1 & \text{if } Q(\mathbf{x}) > k, \\ \delta & \text{if } Q(\mathbf{x}) = k, \\ 0 & \text{if } Q(\mathbf{x}) < k, \end{cases}$$

where k and δ are determined so that $E_{\theta_0}[\phi(\mathbf{X})] = P_{\theta_0}[Q(\mathbf{X}) > k] + \delta P_{\theta_0}[Q(\mathbf{X}) = k] = \alpha$. Also $E_\theta[\phi(\mathbf{X})]$ ($\leqslant 1$) is a nondecreasing function of θ. For the test of hypothesis $H_0: \theta \in \omega \equiv \{\theta \in \Omega; \theta \geqslant \theta_0\}$ against $H_A: \theta \in \Omega - \omega$, a UMP size α-test within the class of all size $\leqslant \alpha$-tests is given by

$$\phi(\mathbf{x}) = \begin{cases} 1 & \text{if } Q(\mathbf{x}) < k, \\ \delta & \text{if } Q(\mathbf{x}) = k, \\ 0 & \text{if } Q(\mathbf{x}) > k, \end{cases}$$

where k and δ are determined so that

$$E_{\theta_0}[\phi(\mathbf{X})] = P_{\theta_0}[Q(\mathbf{X}) < k] + \delta P_{\theta_0}[Q(\mathbf{X}) = k] = \alpha.$$

Also $E_\theta[\phi(\mathbf{X})]$ ($\leqslant 1$) is a nondecreasing function of θ.

(iv) Let $X_1, \ldots, X_n$ be independent identically distributed random variables each with a probability density or probability mass function of the exponential family type (§3.32(I)): $f(x; \theta) = a(\theta)\exp[\theta t(x)]h(x)$. Consider the test of hypothesis $H_0: \theta \in \omega \equiv \{\theta \in \Omega: \theta \leqslant \theta_1 \text{ or } \theta \geqslant \theta_2\}$ against $H_A: \theta \in \Omega - \omega$, with θ_1, θ_2 fixed in Ω, and $\theta_1 < \theta_2$. Then a UMP size α-test, $0 < \alpha < 1$, is given by

$$\phi(\mathbf{x}) = \begin{cases} 1 & \text{if } k_1 < \mathrm{T}(\mathbf{x}) < k_2, \\ \delta_i & \text{if } T(\mathbf{x}) = k_i, \quad i = 1, 2\ (k_1 < k_2), \\ 0 & \text{otherwise}, \end{cases}$$

where $T(\mathbf{x}) = \sum_{i=1}^n t(x_i)$, and k_1, k_2, δ_1, δ_2 are determined so that $E_{\theta_i}[\phi(\mathbf{X})] = \alpha$ for $i = 1, 2$. Also $E_\theta[\phi(\mathbf{X})] \geqslant \alpha$ for $\theta \in \Omega - \omega$, that is the test is unbiased.

(v) In (iv) consider instead the test of hypothesis $H_0: \theta \in \omega \equiv \{\theta \in \Omega, \theta_1 \leqslant \theta \leqslant \theta_2\}$ against $H_A: \theta \in \Omega - \omega$, with fixed θ_1, $\theta_2 \in \Omega$. Then for $0 < \alpha < 1$, a *uniformly most powerful unbiased* (UMPU) exists and is given by

$$\phi(\mathbf{x}) = \begin{cases} 1 & \text{if } T(\mathbf{x}) < k_1 \text{ or } T(\mathbf{x}) > k_2, \\ \delta_i & \text{if } T(\mathbf{x}) = k_i, \quad i = 1, 2\ (k_1 < k_2), \\ 0 & \text{otherwise,} \end{cases}$$

where $T(\mathbf{x}) = \sum_{i=1}^{n} t(x_i)$, and $k_1, k_2, \delta_1, \delta_2$ are determined so that

$$E_{\theta_1}[\phi(\mathbf{X})] = \alpha, \qquad E_{\theta_2}[(\phi(\mathbf{X})] = \alpha$$

for $\theta_1 < \theta_2$, and

$$E_{\theta_0}[\phi(\mathbf{X})] = \alpha, \qquad E_{\theta_0}[T(\mathbf{X})\phi(\mathbf{X})] = \alpha E_{\theta_0}[\phi(\mathbf{X})],$$

for $\theta_1 = \theta_2 \equiv \theta_0$. [By definition of unbiasedness $E_\theta[\phi(\mathbf{X})] \geqslant \alpha$ for $\theta \in \Omega - \omega$.] We also note that no UMP test exists which maximizes the power for all $\theta \in \Omega - \omega$. But if the class of tests is restricted to those which are unbiased, then a UMP test exists within this class as given above.

[Cf., Roussas (1973), Bickel and Doksum (1977), Fourgeaud and Fuchs (1967), Lehmann (1959).]

§1.33. Locally Most Powerful Tests

Consider a test of hypothesis on a parameter θ: H_0: $\theta \leqslant \theta_0$ against H_A: $\theta > \theta_0$. A size α-test ϕ^* is said to be locally most powerful (LMP), if, given any other size α-test ϕ, we may find an $\varepsilon > 0$, such that for all θ in $0 < \theta - \theta_0 < \varepsilon$, $E_\theta[\phi^*] \geqslant E_\theta[\phi]$. Quite generally, let ρ denote a measure of distance between an alternative from a null hypothesis ($\rho = \theta - \theta_0 > 0$ for the above test), then a size α-test ϕ^* is locally most powerful if, given any other size α-test ϕ, we may find an $\varepsilon > 0$, such that for all θ *with* $0 < \rho < \varepsilon$, $E_\theta[\phi^*] \geqslant E_\theta[\phi]$.

To determine the locally most powerful test ϕ^* for the test of hypothesis H_0: $\theta \leqslant \theta_0$ against H_A: $\theta > \theta_0$, suppose that $E_\theta[\phi]$ is continuously differentiable at $\theta = \theta_0$ (that may be passed under the expectation value) for every test ϕ. Let $L(\theta; \mathbf{X})$ denote the likelihood function. Then ϕ^* is given by

$$\phi^* \equiv \begin{cases} 1 & \text{if } (\partial/\partial\theta) \ln L(\theta; \mathbf{X})|_{\theta=\theta_0} > k \\ \gamma & \text{if} \qquad\qquad\qquad\qquad = \\ 0 & \text{if} \qquad\qquad\qquad\qquad < \end{cases}, \tag{$*$}$$

where $0 \leqslant \gamma \leqslant 1$, and k are determined so that $E_{\theta_0}[\phi^*] = \alpha$. For the test of hypothesis H_0: $\theta \geqslant \theta_0$ against H_A: $\theta < \theta_0$, simply reverse the inequalities in ($*$). In this case, we note that we may write $\rho = \theta_0 - \theta$. See also §1.44.

[Cf., Lehmann (1959), Ferguson (1967).]

§1.34. Locally Most Powerful Unbiased Tests

Consider a test of hypothesis on a parameter θ, and let ρ denote a measure of distance between an alternative from a null hypothesis, then a size α-test ϕ^* is said to be locally most powerful unbiased test if, given any unbiased size

α-test ϕ, we may find an $\varepsilon > 0$, such that for all θ *with* $0 < \rho < \varepsilon$, $E_\theta[\phi^*] \geqslant E_\theta[\phi]$. Consider the test of hypothesis $H_0: \theta = \theta_0$ against $H_A: \theta \neq \theta_0$. We may take $\rho = |\theta - \theta_0|$. Suppose that $E_\theta[\phi]$ is twice continuously differentiable at $\theta = \theta_0$ (that may be passed under the expectation value) for every ϕ. Then ϕ^* is given by

$$\phi^* = \begin{cases} 1 & \text{if } [(\partial^2/\partial\theta^2)\ln L(\theta;\mathbf{X})]_{\theta=\theta_0} + [(\partial/\partial\theta)\ln L(\theta;\mathbf{X})]^2_{\theta=\theta_0}, \\ & \qquad > k_1 + k_2[(\partial/\partial\theta)\ln L(\theta;\mathbf{X})]_{\theta=\theta_0}, \\ \gamma & \text{if} \qquad = \\ 0 & \text{if} \qquad < \end{cases}$$

where $0 \leqslant \gamma \leqslant 1$, k_1, k_2 are determined so that $E_{\theta_0}[\phi^*] = \alpha$, $(\partial/\partial\theta)E_\theta[\phi^*]|_{\theta=\theta_0} = 0$, and where $L(\theta; \mathbf{X})$ denotes the likelihood function.

[Cf., Lehmann (1959), Ferguson (1967).]

§1.35. Likelihood Ratio Test

Let $X_1, \ldots, X_n$ be independent identically distributed random variables each with a probability density or probability mass $f(x; \theta)$, $\theta \in \Omega$. Consider the test of hypothesis $H_0: \theta \in \omega$ against $H_A: \theta \in \Omega - \omega$. Define the likelihood $L(\theta; \mathbf{x}) = \prod_{i=1}^n f(x_i; \theta)$. The likelihood ratio statistic is defined by $\Lambda_n = \sup_{\theta\in\omega} L(\theta; \mathbf{x}) / \sup_{\theta\in\Omega} L(\theta; \mathbf{x})$. [Note that $0 \leqslant \Lambda_n \leqslant 1$.] Then with a level of significance α, H_0 is rejected if $\Lambda_n \leqslant \lambda_n(\alpha)$, where $\lambda_n(\alpha)$ is determined so that $\sup_{\theta\in\omega} P[\Lambda_n \leqslant \lambda_n(\alpha)] = \alpha$.

For the asymptotic distribution of Λ_n and consistency of the test for $\omega = \{\theta_0\}$, see §2.26. The likelihood ratio test provides a unified and intuitively appealing method for developing tests of hypotheses. It often produces a test which is "close" to a UMP test. It may be applied in the presence of several parameters and for composite hypotheses as well. It also has asymptotic (§2.26) optimum properties.

As an example, consider the test of homogeneity of variances $\sigma_1^2 = \cdots = \sigma_k^2$ associated with k *normal* populations. The likelihood ratio statistic Λ_n then becomes

$$\Lambda_n = \prod_{i=1}^k [\tilde{S}_i^2/\tilde{S}^2]^{n_i/2}, \quad \text{where} \quad \tilde{S}_i^2 = \sum_{j=1}^{n_i} (X_{ij} - \bar{X}_{i.})^2/n_i,$$

$$\tilde{S}^2 = \sum_{i=1}^k n_i \tilde{S}_i^2/n, \qquad n = \sum_{i=1}^k n_i,$$

$X_{i1}, \ldots, X_{in_i}$ denotes the ith sample of size n_i taken from the ith population, and $\bar{X}_{i.}$ is the ith sample mean. The likelihood ratio test then rejects the hypothesis $H_0: \sigma_1^2 = \cdots = \sigma_k^2$, against H_A: not all the σ_i^2 are equal, for large values of the statistic $\tilde{M} = -\sum_{i=1}^k n_i \ln \tilde{S}_i^2 + n \ln \tilde{S}^2$. Unless all the n_i are equal the test is generally biased. To make the test unbiased Bartlett replaced [cf. Kendall and Stuart (1979, pp. 252, 261)] the n_i by the "degrees of freedom" $\nu_i = n_i - 1$ obtaining the statistic

$$M = -\sum_{i=1}^{k} \nu_i \ln S_i^2 + \nu \ln S^2,$$

where $S_i^2 = \sum_{j=1}^{k} (X_{ij} - \bar{X}_{i.})^2/(n_i - 1)$, $S^2 = \sum_{i=1}^{k} \nu_i S_i^2/\nu$, $\nu = \sum_{i=1}^{k} \nu_i$. The resulting test is then unbiased (also consistent) and rejects H_0 for large values of the statistic. For details on the distribution of M see §§2.15(v) and 3.48. If the normality condition is in doubt, it is not advisable to use Bartlett's test for its lack of robustness (see §1.42) to departures from normality. In this case one may use the robust jackknife method described in §1.42.

[Cf., Lehmann (1959), Bickel and Doksum (1977), Roussas (1973), Fourgeaud and Fuchs (1967), Kendall and Stuart (1979).]

§1.36. Theorems on Unbiasedness of Tests

Theorem 1 (Lehmann). *Let $X_1, \ldots, X_n$ be a sample from a continuous distribution $F(x)$, and consider the test of hypothesis H_0: $F(x) = F_0(x)$ for all x, against H_A: $F(x) < F_0(x)$ for at least one x. Suppose, that with a level of significance α; H_0 is rejected for $T_n(X_1, \ldots, X_n) \geqslant t_n$, where $T_n(X_1, \ldots, X_n)$ is the statistic in question. If $T_n(x_1, \ldots, x_n) \geqslant T_n(x_1', \ldots, x_n')$ for every $x_1 \geqslant x_1', \ldots, x_n \geqslant x_n'$, then $P_F[T_n \geqslant t_n] \geqslant P_{F_0}[T_n \geqslant t_n]$ under H_A, that is for $F(x) < F_0(x)$ (for at least one x).*

Theorem 2 (Lehmann). *Let $X_{11}, \ldots, X_{1n_1}$ and $X_{21}, \ldots, X_{2n_2}$ be two independent samples from continuous distributions $F_1(x)$ and $F_2(x)$, respectively, and consider the test of hypothesis H_0: $F_1(x) = F_2(x)$ for all x, against H_A: $F_1(x) < F_2(x)$ for at least one x. Suppose that, with a level of significance α, H_0 is rejected for $T_{n_1,n_2}(X_{11}, \ldots, X_{1n_1}; X_{21}, \ldots, X_{2n_2}) \geqslant t_{n_1,n_2}$, where $T_{n_1,n_2}(X_{11}, \ldots, X_{1n_1}; X_{21}, \ldots, X_{2n_2})$ is the statistic in question. If*

$$T_{n_1,n_2}(x_{11}, \ldots, x_{1n_1}; x_{21}, \ldots, x_{2n_2}) \geqslant T_{n_1,n_2}(x_{11}', \ldots, x_{1n_1}'; x_{21}, \ldots, x_{2n_2})$$

for every $x_{11} \geqslant x_{11}', \ldots, x_{1n_1} \geqslant x_{1n_1}'$, then

$$P_{F_1,F_2}[T_{n_1,n_2} \geqslant t_n] \geqslant P_{F_2,F_2}[T_{n_1,n_2} \geqslant t_n]$$

under H_A, that is for $F_1(x) < F_2(x)$ (for at least one x).

[Cf., Randles and Wolfe (1979), Lehmann (1951).]

§1.37. Relative Efficiency of Tests

One may compare the performance of a test statistic $T^{(1)}$ to a test statistic $T^{(2)}$, for a given test of hypothesis, by comparing the sample sizes needed for each of these test statistics to achieve the same given power. That is, at a given level of significance α, if the $T^{(1)}$-test requires n_1 observations to achieve a certain given power, and if the $T^{(2)}$-test requires n_2 observations to achieve this same power, then the relative efficiency of the $T^{(1)}$-test to the $T^{(2)}$-test may be defined by $e(T^{(1)}, T^{(2)}) = n_2/n_1$. If $e(T^{(1)}, T^{(2)}) > 1$, then the $T^{(1)}$-test may be termed as more efficient than the $T^{(2)}$-test as it requires less observations than the latter to achieve the given power in question. And the $T^{(1)}$-test is, in general, more sensitive to deviations from the null hypothesis. One should, however, be careful with such interpretations as the value of $e(T^{(1)}, T^{(2)})$, in general, depends on the given size α of the tests, the given power of the test, and the "given degree of violation" of the null hypothesis, and may have values less than one for some cases and values greater than one for other cases. For an *asymptotic* relative efficiency of tests, see §2.28 (Pitman Asymptotic Efficiency), which gives a simple expression or leads to a fixed number, as a measure of relative efficiency of tests.

[See also, for example, Serfling (1980) for other measures of efficiency.]

§1.38. Sequential Probability Ratio Test (SPRT)

Let $X_1, X_2, \ldots$ be independent identically distributed random variables each with probability density or probability mass $f(x)$. We are interested in testing the hypothesis H_0: $f \equiv f_0$ against H_A: $f \equiv f_1$ *without fixing in advance the sample size n* (the sequential procedure).

Define $Z_i = \ln[f_1(X_i)/f_0(X_i)]$, and for each n, set

$$S_n = \sum_{i=1}^{n} Z_i, \qquad \Lambda_n = \prod_{i=1}^{n} [f_1(x_i)/f_0(x_i)].$$

Suppose $P_i[Z_1 = 0] \neq 1$, that is, $P_i[f_0(X) = f_1(X)] \neq 1$, for $i = 0, 1$. The sequential procedure for the test of hypothesis H_0 against H_A proceeds as follows. Let A and B be two fixed numbers $0 < A < 1 < B$. Continue sampling as long as $A < \Lambda_n < B$. As soon as a sample size n is obtained such that $\Lambda_n \leqslant A$ then accept H_0 and stop sampling, and if $\Lambda_n \geqslant B$, then reject H_0 and stop sampling. The constants A and B may be chosen to obtain (approximately) the assigned Type-I α and Type-II β errors, where $0 < \alpha < \beta < 1$, $0 < \alpha + \beta < 1$. Let N denote the random variable for the number of observations to be taken in the sequential procedure. Then

(i) There exists a $\delta: 0 < \delta < 1$, and a constant $c > 0$ such that for all n, $P_i[N \geqslant n] \geqslant c\delta^n$, $P_i[N < \infty] = 1$, $E_i[N] < \infty$ for $i = 0, 1$.

(ii) $A \geqslant \beta/(1 - \alpha)$, $B \leqslant (1 - \beta)/\alpha$. In practice one makes the approximation $A \approx \beta/(1 - \alpha)$, $B \approx (1 - \beta)/\alpha$ which is often justifiable.

(iii) Among all tests for which P_0[rejecting H_0] $\leqslant \alpha$, P_1[accepting H_0] $\leqslant \beta$, the sequential probability ratio test (SPRT) minimizes both $E_0[N]$ and $E_1[N]$.

(iv) (Wald's Equations). If $E_i[|Z_1|] < \infty$, then $E_i[|S_N|] < \infty$ and $E_i[S_N] = E_i[N]E_i[Z_1]$, for $i = 0, 1$.

(v) Wald's equations provide the following practical approximations for the average sample sizes in the SPRT: $E_0[N] \approx (\alpha \ln B + (1-\alpha)\ln A)/E_0[Z_1]$, $E_1[N] \approx ((1-\beta)\ln B + \beta \ln A)/E_1[Z_1]$.

(vi) Let $f_0(x) \equiv f(x; \theta_0)$, $f_1(x) \equiv f(x; \theta_1)$, $\theta_1 > \theta_0$, and consider the test of hypothesis $H_0: \theta = \theta_0$ against $H_A: \theta = \theta_1$. Suppose that $f(x; \theta)$ has a monotone likelihood ratio property in $Q(x)$ (with the latter independent of θ): that is, $f(x; \theta) \neq f(x; \theta')$ and $f(x; \theta')/f(x; \theta)$ $(\theta' > \theta)$ is a strictly increasing function in Q for all x in the set of positivity of $f(x; \theta)$: $\{x: f(x; \theta) > 0\}$ which is assumed to be independent of θ. (Lehmann): Then the power is a nondecreasing function of θ.

(vii) (Wald). If $E_\theta[(f(X; \theta_1)/f(X; \theta_0))^h] = 1$ has a nonzero solution h, then as an approximate expression, the power $\approx [1 - (A)^h]/[(B)^h - (A)^h]$. [Sufficiency conditions (Lehmann) for the existence of a nonzero solution h are: $E_\theta[Z] \neq 0$, $E_\theta[e^{hZ}]$ exists for all h, and $P_\theta[e^Z < 1 - \delta] > 0$, $P_\theta[e^Z > 1 + \delta] > 0$ for some $\delta > 0$, where $Z = \ln[f(X; \theta_1)/f(X; \theta_0)]$.]

(viii) As an application to the binomial distribution consider the test of hypothesis: $H_0: p = p_0$ against $H_A: p = p_1 (> p_0)$. Then as long as the number of "successes" $X(n)$ is such that $A_0 + Bn < X(n) < A_1 + Bn$, where

$$A_0 = \frac{\ln[\beta/(1-\alpha)]}{\ln[p_1(1-p_0)/p_0(1-p_1)]},$$

$$A_1 = \frac{\ln[(1-\beta)/\alpha]}{\ln[p_1(1-p_0)/p_0(1-p_1)]},$$

$$B = \frac{\ln[(1-p_0)/(1-p_1)]}{\ln[p_1(1-p_0)/p_0(1-p_1)]},$$

one continues to take observations. After having continued on the process if one reaches a value of n for which $X(n) \geqslant A_1 + Bn$, H_0 is rejected and the sampling is stopped; on the other hand, if $X(n) \leqslant A_0 + Bn$, H_0 is accepted and the sampling is stopped.

As an application to the normal distribution with mean μ and variance $\sigma^2 = 1$, consider the test of hypothesis: $H_0: \mu = \mu_0$ against $H_A: \mu = \mu_1$ $(> \mu_0)$. If $X_1, \ldots, X_n$ denotes a sample and $X(n) = X_1 + \cdots + X_n$, then continue sampling as long as $A_0 + Bn < X(n) < A_1 + Bn$, where

$$A_0 = \frac{\ln[\beta/(1-\alpha)]}{(\mu_1 - \mu_0)},$$

$$A_1 = \frac{\ln[(1-\beta)/\alpha]}{(\mu_1 - \mu_0)},$$

$$B = \frac{(\mu_1 + \mu_0)}{2}.$$

After having continued on the process if one reaches a value of n for which $X(n) \geqslant A_1 + Bn$, H_0 is rejected and the sampling is stopped; on the other hand, if $X(n) \leqslant A_0 + Bn$, H_0 is accepted and the sampling is stopped.

[Cf., Wald (1947), Roussas (1973), Wetherill (1980), Silvey (1975).]

§1.39. Bayes and Decision-Theoretic Approach

(i) Let $X_1, \ldots, X_n$ be independent identically distributed random variables each with probability density or probability mass function $f(x; \theta)$, $\theta \in \Omega$, where Ω is the parameter space and is often referred to as the "*state space*," and θ specifies the state of Nature.

A *decision function* $\delta(x_1, \ldots, x_n) \equiv \delta(\mathbf{x})$ is a real function on R^n. A *loss function* is a nonnegative function of θ and δ: $l(\theta, \delta)$ and expresses the loss incurred when θ is estimated by $\delta(\mathbf{x})$, where $x_1, \ldots, x_n$ are the values taken by the sample, or more generally if the decision function $\delta(\mathbf{x})$ is used to reach a conclusion on θ when the latter is the true state of Nature.

The risk function corresponding to the decision function δ is defined by $R(\theta, \delta) = E_\theta[l(\theta, \delta(\mathbf{X}))]$, where the expectation is taken with respect to the distribution of $\mathbf{X} = (X_1, \ldots, X_n)'$.

Typically, if $\delta(\mathbf{X})$ is an unbiased estimator of θ, and $l(\theta, \delta)$ is taken to be the squared loss function: $l(\theta, \delta) = (\delta(\mathbf{x}) - \theta)^2$, then the risk function $R(\theta, \delta)$ is simply the variance of $\delta(\mathbf{x})$ (assuming it exists).

On the other hand, consider the test of hypothesis $H_0\colon \theta \in \omega$ against $H_A\colon \theta \in \Omega - \omega$, and suppose the decision function $\delta(\mathbf{x})$ concerning θ may take on only the values 0 or 1, and if $\delta(\mathbf{x}) = 1$, H_0 is rejected, and if $\delta(\mathbf{x}) = 0$, H_0 is accepted. That is, $\delta(\mathbf{x})$ is a (nonrandomized) test function. Define the loss function:

$$l(\theta, \delta) = \begin{cases} 0 & \text{if } \delta = 1, \quad \theta \in \Omega - \omega, \\ 0 & \text{if } \delta = 0, \quad \theta \in \omega, \\ 1 & \text{if } \delta = 1, \quad \theta \in \omega, \\ 1 & \text{if } \delta = 0, \quad \theta \in \Omega - \omega. \end{cases}$$

Hence for $\theta \in \omega$, $R(\theta, \delta) = P_\theta[\delta(\mathbf{X}) = 1]$ is a Type-I error, and for $\theta \in \Omega - \omega$, $R(\theta, \delta) = P_\theta[\delta(\mathbf{X}) = 0]$ is a Type-II error.

A decision function $\delta(\mathbf{x})$ is called *admissible* if there exists no other decision function $\tilde{\delta}(\mathbf{x})$ such that $R(\theta, \tilde{\delta}) \leqslant R(\theta, \delta)$ with strict inequality holding for at least one $\theta \in \Omega$. Two decision functions δ, $\tilde{\delta}$ are said to be *equivalent* if $R(\theta, \delta) = R(\theta, \tilde{\delta})$ for all $\theta \in \Omega$. A class C of decision functions is said to be *essentially complete* if for any decision function $\tilde{\delta}$ not in C, we may find a decision function δ in C such that $R(\theta, \delta) \leqslant R(\theta, \tilde{\delta})$. Within the essentially complete class of decision functions for which $R(\theta, \delta)$ is finite for all $\theta \in \Omega$, a decision function δ is said to be *minimax* if $\sup_{\theta \in \Omega} R(\theta, \delta) \leqslant \sup_{\theta \in \Omega} R(\theta, \tilde{\delta})$ for any other decision function $\tilde{\delta}$ concerning θ.

(ii) Suppose one assigns a probability distribution for θ in Ω with probability density or probability mass function $\lambda(\theta)$, $\theta \in \Omega$. $\lambda(\theta)$ is called the prior probability density or probability mass function of $\theta \in \Omega$. If a sample $X_1, \ldots, X_n$ gives the values $x_1, \ldots, x_n$, then the conditional probability density function or the conditional probability mass function of θ is given by:

$$\lambda(\theta|\mathbf{x}) = \frac{L(\theta; \mathbf{x})\lambda(\theta)}{E_\lambda[L(\theta; \mathbf{x})]},$$

where the expectation E_λ is taken with respect to $\lambda(\theta)$ of θ over Ω, and $L(\theta; \mathbf{x})$ is the likelihood function: $L(\theta; \mathbf{x}) = \prod_{i=1}^n f(x_i; \theta)$. $\lambda(\theta|\mathbf{x})$ is called the *posterior* probability density or *posterior* probability mass function of θ based on the information obtained from the sample.

The *Bayes risk* is defined by $R(\delta) = E_\lambda[R(\theta, \delta)]$, where $R(\theta, \delta)$ is the risk function introduced in (i). We may also write $R(\delta)$ as a double expectation with respect to $L(\theta; \mathbf{x})\lambda(\theta)$: $R(\delta) = E[l(\theta, \delta(\mathbf{X}))]$. Within the class of all decision functions for which $R(\delta)$ is finite, δ is called a *Bayes rule* with respect to λ if for any other decision functions $\tilde{\delta}$ concerning θ, $R(\delta) \leqslant R(\tilde{\delta})$.

Theorem. *If for each* $\mathbf{x}$, *we may find a decision function* $\delta(\mathbf{x})$ *such that*

$$E_\lambda[l(\theta, \delta(\mathbf{x}))L(\theta; \mathbf{x})] \leqslant E_\lambda[l(\theta, \tilde{\delta}(\mathbf{x}))L(\theta; \mathbf{x})],$$

for any other decision function $\tilde{\delta}$, *concerning* θ, *where the expectation* E_λ *is with respect to* $\lambda(\theta)$, *and the latter expectations are assumed to be finite, then* $\delta(\mathbf{x})$ *is a Bayes rule.*

For example, for the squared loss function $l(\theta, \delta) = (\delta(\mathbf{x}) - \theta)^2$, a Bayes rule is given by

$$\delta(\mathbf{x}) = \frac{E_\lambda[\theta L(\theta; \mathbf{x})]}{E_\lambda[L(\theta; \mathbf{x})]}.$$

Theorem. *If* $\Omega = \{\theta_1, \ldots, \theta_k\}$ *and* $\lambda(\theta_i) > 0$ *for each* $\theta_i \in \Omega$, *and* δ *is a Bayes rule, with respect to* λ, *then* δ *is admissible.*

Theorem. *Suppose* $\Omega = (a, b)$ *(a and b may* $-\infty$ *and* $+\infty$*), and* $\lambda(\theta) > 0$ *for all* $\theta \in \Omega$. *Suppose that* $R(\theta, \delta)$ *is a continuous function of* θ *for all* δ *within the class of decision functions, with respect to* λ, *for which* $R(\delta) < \infty$. *If* δ *is a Bayes rule, with respect to* λ, *then* δ *is admissible.*

Theorem. *Suppose* δ *is a Bayes rule with respect to* $\lambda(\theta)$ *on a state space* Ω. *If* $R(\theta, \delta)$ *is independent of* θ *for all* $\theta \in \Omega$, *then* δ *is minimax.*

Although a uniformly minimum variance unbiased estimator of a parameter θ has an obvious optimal property, in some cases it may yield an absurd result about θ. For example, one may obtain a negative number in estimating a positive parameter. A Bayes estimator (rule), although not altogether void of arbitrariness in choosing a prior probability distribution, has an advantage in that such difficulties may, in general, be avoided by a proper choice of $\lambda(\theta)$.

Finally, we consider an intimate connection that exists between the Bayes method and the Neyman–Pearson lemma. Consider the test of hypothesis H_0: $\theta = \theta_0$ against H_A: $\theta = \theta_1, \Omega = \{\theta_0, \theta_1\}$. Define the decision function $\delta(\mathbf{x})$ taking the values 0 or 1, and if $\delta(\mathbf{x}) = 1$ reject H_0, and if $\delta(\mathbf{x}) = 0$ accept H_0. Define the loss function:

$$l(\theta, \delta) = \begin{cases} 0 & \text{if } \delta = 1, \quad \theta = \theta_1, \\ 0 & \text{if } \delta = 0, \quad \theta = \theta_0, \\ l_0 & \text{if } \delta = 1, \quad \theta = \theta_0, \\ l_1 & \text{if } \delta = 0, \quad \theta = \theta_1, \end{cases}$$

where l_0 and l_1, for example, may have units of dollars. Suppose $\lambda(\theta)$ denotes the probability mass function of θ. If a sample $(X_1, \ldots, X_n)$ gives the values $(x_1, \ldots, x_n)' \equiv \mathbf{x}$, then the posterior probability mass function of θ is

$$\lambda(\theta|\mathbf{x}) = \frac{L(\theta; \mathbf{x})\lambda(\theta)}{\sum_{i=1}^{2} [L(\theta_i; \mathbf{x})\lambda(\theta_i)]}, \quad \text{where} \quad L(\theta; \mathbf{x}) = \prod_{i=1}^{n} f(x_i; \theta).$$

Accordingly, given that the sample is x, the (posterior) expected losses in rejecting H_0 and accepting H_0 are, respectively, $E_\lambda[l(\theta, \delta = 1)] = l_0\lambda(\theta_0|\mathbf{x})$, $E_\lambda[l(\theta), \delta = 0] = l_1\lambda(\theta_1|\mathbf{x})$. Hence, if $l_1\lambda(\theta_1|\mathbf{x}) > l_0\lambda(\theta_0|\mathbf{x})$, then we reject H_0, since the corresponding posterior expected loss is smaller than the corresponding one if we accept H_0. The rejection region, in the Bayesian method, then becomes $\{\mathbf{x}: L(\theta_1; \mathbf{x}) > kL(\theta_0; \mathbf{x})\}$, where $k = l_0\lambda(\theta_0)/l_1\lambda(\theta_1)$. This should be compared with the rejection region given in the Neyman–Pearson fundamental lemma (for a nonrandomized test). Accordingly, given k, one may obtain the size α of the test in the classical approach, and the Bayesian approach may give some guideline on how to choose α. Conversely, given α, l_0 and l_1, one may assign values $\lambda(\theta_0)$ and $\lambda(\theta_1)$, thus defining a prior probability for θ. For $L(\theta_1; \mathbf{x}) = kL(\theta_0; \mathbf{x})$, one may choose either decisions in the Bayesian approach, or toss a coin to decide which hypothesis to choose. The latter is, of course, in the same spirit as a randomization test in the classical approach.

[Cf., Berger (1980), Box and Tiao (1973), Ferguson (1967), Hartigan (1983), Roussas (1973), Silvey (1975), Bickel and Doksum (1977), Cox and Hinkley (1974), Beaumont (1980).]

§1.40. The Linear Hypothesis

The general linear model is defined by $\mathbf{Y} = \mathbf{X}\boldsymbol{\beta} + \boldsymbol{\varepsilon}$, where $\mathbf{Y} = (Y_1, \ldots, Y_n)'$ is a random n-vector, $\mathbf{X}$ is a $n \times p$ known matrix ($p \leqslant n$), $\beta = (\beta_1, \ldots, \beta_p)'$ is a p-vector with unknown components and is called the vector parameter, and

$\boldsymbol{\varepsilon} = (\varepsilon_1, \ldots, \varepsilon_n)'$, called the error, is a random n-vector with zero mean: $E[\varepsilon_i] = 0$, that is $E[\boldsymbol{\varepsilon}] = 0$. We will also assume that $0 < \sigma^2(\varepsilon_i) = \sigma^2 < \infty$ for $i = 1, \ldots, n$, and that the ε_i are not correlated, that is, $\text{Cov}[\varepsilon_i, \varepsilon_j] = 0$, for $i \neq j$. It is called a linear model because the Y_i depend on the parameters $\beta_1, \ldots, \beta_p$ linearly:

$$Y_i = X_{i1}\beta_1 + \cdots + X_{ip}\beta_p + \varepsilon_i,$$

where X_{ij} are the matrix elements of $\mathbf{X}$. We note that $E[\mathbf{Y}] = \mathbf{X}\boldsymbol{\beta} \equiv \boldsymbol{\theta}$.

The *least-squares estimate* of $\boldsymbol{\beta}$ is defined by minimizing the expression $\boldsymbol{\varepsilon}'\boldsymbol{\varepsilon} = \sum_{i=1}^{n} \varepsilon_i^2$ with respect to the β_i. The expression $\boldsymbol{\varepsilon}'\boldsymbol{\varepsilon}$ may be written as $\boldsymbol{\varepsilon}'\boldsymbol{\varepsilon} = (\mathbf{Y} - \mathbf{X}\boldsymbol{\beta})'(\mathbf{Y} - \mathbf{X}\boldsymbol{\beta})$. Any solution which minimizes the expression $\boldsymbol{\varepsilon}'\boldsymbol{\varepsilon}$ will be denoted by $\hat{\boldsymbol{\beta}}$.

Theorem. *Any solution of the equations* $\mathbf{X}'\mathbf{X}\boldsymbol{\beta} = \mathbf{X}'\mathbf{Y}$ *is a least-squares estimate of* β, *and a least-squares estimate* $\hat{\boldsymbol{\beta}}$ *of* $\boldsymbol{\beta}$ *is a solution of the above equations. These equations are called the* normal equations.

If rank $\mathbf{X} = p$, then $\mathbf{X}'\mathbf{X}$ has an inverse and the normal equations admit a unique solution $\hat{\boldsymbol{\beta}} = (\mathbf{X}'\mathbf{X})^{-1}\mathbf{X}'\mathbf{Y}$, $E[\hat{\boldsymbol{\beta}}] = \boldsymbol{\beta}$, that is, in particular, $\hat{\boldsymbol{\beta}}$ is an unbiased estimator of $\boldsymbol{\beta}$. Also in this case, $\text{Cov}[\hat{\beta}_i, \hat{\beta}_j] = 0$, for $i \neq j$, $\sigma^2(\hat{\beta}_i) = \sigma^2(\mathbf{X}'\mathbf{X})^{-1}$. If rank $\mathbf{X} < p$, then no unique solution exists and $\boldsymbol{\beta}$ is called unidentifiable.

Let rank $\mathbf{X} = r$. Since $E[\mathbf{Y}] = \mathbf{X}\boldsymbol{\beta} \equiv \boldsymbol{\theta}$,

$$(\mathbf{X}\boldsymbol{\beta})_i = X_{i1}\beta_1 + \cdots + X_{ip}\beta_p,$$

we note that $\boldsymbol{\theta}$ is a vector which lies in a subspace Ω_r generated by all linear combinations of the columns of the matrix $\mathbf{X}$. That is, any vector in Ω_r may be written as a linear combination of the column vectors of $\mathbf{X}$. By definition of the rank of a matrix, Ω_r is of r dimensions. Ω_r is a subspace of some n-dimensional vector space which will be denoted by V_n: $\Omega_r \subseteq V_n$. If we set $\hat{\boldsymbol{\theta}} = \mathbf{X}\hat{\boldsymbol{\beta}}$, where $\hat{\boldsymbol{\beta}}$ is a least-squares estimate of $\boldsymbol{\beta}$, then we note from

$$\mathbf{X}'(\mathbf{Y} - \hat{\boldsymbol{\theta}}) = \mathbf{X}'\mathbf{Y} - \mathbf{X}'\hat{\boldsymbol{\theta}} = \mathbf{X}'\mathbf{Y} - \mathbf{X}'\mathbf{X}\hat{\boldsymbol{\beta}} = \mathbf{X}'\mathbf{Y} - \mathbf{X}'\mathbf{Y} = 0,$$

where we have used the normal equations, that $\mathbf{Y} - \hat{\boldsymbol{\theta}}$ is orthogonal to every vector in Ω_r. We may write $\mathbf{Y} = \mathbf{Y}_{\parallel} + \mathbf{Y}_{\perp}$, where $\mathbf{Y}_{\parallel} \in \Omega_r$, and $\mathbf{Y}_{\perp}$ is orthogonal to $\mathbf{Y}_{\parallel}$. Since $\hat{\boldsymbol{\theta}} \in \Omega_r$, $\mathbf{Y} - \hat{\boldsymbol{\theta}} = (\mathbf{Y}_{\parallel} - \hat{\boldsymbol{\theta}}) + \mathbf{Y}_{\perp}$, and by the fact that $\mathbf{Y} - \hat{\boldsymbol{\theta}}$ is orthogonal to Ω_r, we arrive to the conclusion that $\mathbf{Y}_{\parallel} - \hat{\boldsymbol{\theta}} = 0$. That is, $\boldsymbol{\theta}$ is estimated by a vector $\hat{\boldsymbol{\theta}}$ in Ω_r which is of closest distance from $\mathbf{Y}$.

Definition. Let $\mathbf{w}$ be a given p-vector. Then $\mathbf{w}'\boldsymbol{\beta}$ is called *estimable* if there exists a vector $\mathbf{u} \in V_n$ such that $\mathbf{u}'\mathbf{Y}$ is an unbiased estimator of $\mathbf{w}'\boldsymbol{\beta}$: $E[\mathbf{u}'\mathbf{Y}] = \mathbf{w}'\boldsymbol{\beta}$ *identically* in $\boldsymbol{\beta}$. $\mathbf{w}'\boldsymbol{\beta}$ is called a linear parametric function.

We note that if $\mathbf{w}'\boldsymbol{\beta}$ is estimable, then from $E[\mathbf{Y}] = \mathbf{X}\boldsymbol{\beta}$, we infer that $\mathbf{u}'\mathbf{X}\boldsymbol{\beta} = \mathbf{w}'\boldsymbol{\beta}$ and hence $\mathbf{u}'\mathbf{X} = \mathbf{w}'$, since the former relation must hold identically in $\boldsymbol{\beta}$, by definition. Also if we write $\mathbf{u}' = \mathbf{u}'_{\parallel} + \mathbf{u}'_{\perp}$, using the previous notation, we note that $\mathbf{u}'_{\parallel}\mathbf{Y}$ is also an unbiased estimator of the estimable linear parametric function $\mathbf{w}'\boldsymbol{\beta}$ since $\mathbf{u}'_{\perp}\mathbf{X} = 0$. Also if $\hat{\boldsymbol{\beta}}$ is a least-squares estimator of $\boldsymbol{\beta}$, then from the relation $\mathbf{u}'_{\parallel}\mathbf{Y} - \mathbf{w}'\hat{\boldsymbol{\beta}} = \mathbf{u}'_{\parallel}(\mathbf{Y} - \mathbf{X}\hat{\boldsymbol{\beta}}) = 0$, where we have used the identity $\mathbf{u}'\mathbf{X} = \mathbf{w}'$, for an estimable $\mathbf{w}'\boldsymbol{\beta}$, and the fact that $(\mathbf{Y} - \mathbf{X}\hat{\boldsymbol{\beta}})$ is orthogonal to every vector in Ω_r, and hence also to $\mathbf{u}'_{\parallel}$. Thus we arrive at the conclusion that $\mathbf{u}'_{\parallel}\mathbf{Y} = \mathbf{w}'\hat{\boldsymbol{\beta}}$, for an estimable $\mathbf{w}'\boldsymbol{\beta}$, where $\hat{\boldsymbol{\beta}}$ is a least-squares estimator of $\boldsymbol{\beta}$. In particular we note, by definition, that if rank $\mathbf{X} = p$, then $\mathbf{w}'\boldsymbol{\beta}$ is estimable for every p-vector $\mathbf{w} \in V_n$, with $\mathbf{u}' = \mathbf{w}'(\mathbf{X}'\mathbf{X})^{-1}\mathbf{X}'$.

The Gauss–Markov Theorem. *Consider the linear model*:

$$\mathbf{Y} = \mathbf{X}\boldsymbol{\beta} + \boldsymbol{\varepsilon}, \qquad X = [X_{ij}]_{n\times p}, \qquad p \leqslant n,$$

$$\sigma^2(\varepsilon_i) = \sigma^2 < \infty, \quad i = 1, \ldots, n, \qquad \operatorname{Cov}[\varepsilon_i, \varepsilon_j] = 0 \quad \text{for } i \neq j.$$

For every p-vector $\mathbf{w}$ *such that* $\mathbf{w}'\boldsymbol{\beta}$ *is estimable,* $\mathbf{w}'\hat{\boldsymbol{\beta}}$ *is an unbiased estimator of* $\mathbf{w}'\boldsymbol{\beta}$, *and* $\sigma^2(\mathbf{w}'\hat{\boldsymbol{\beta}}) \leqslant \sigma^2(\mathbf{w}'\tilde{\boldsymbol{\beta}})$, *where* $\hat{\boldsymbol{\beta}}$ *is a least-squares estimator of* $\boldsymbol{\beta}$, *and* $\tilde{\boldsymbol{\beta}}$ *is any other linear (in* $\mathbf{Y}$, *i.e.,* $\tilde{\boldsymbol{\beta}} = \mathbf{c}'\mathbf{Y}$) *unbiased estimator of* $\boldsymbol{\beta}$. *That is, in particular,* $\mathbf{w}'\hat{\boldsymbol{\beta}}$ *has the smallest variance in the class of all linear unbiased estimators* $\mathbf{w}'\tilde{\boldsymbol{\beta}}$ *of* $\mathbf{w}'\boldsymbol{\beta}$.

In particular, we note that if $\mathbf{w} = (0, \ldots, 0, 1, 0, \ldots, 0)'$, with 1 at the ith place, then $\hat{\beta}_i$ has the smallest variance in the class of all linear unbiased estimators of β_i.

Theorem. *Let rank* $\mathbf{X} = r < n$, *then an unbiased estimator of* σ^2 *is provided by* $\hat{\sigma}^2 = (\mathbf{Y} - \mathbf{X}\hat{\boldsymbol{\beta}})'(\mathbf{Y} - \mathbf{X}\hat{\boldsymbol{\beta}})/(n - r)$. *If in addition,* $r = p$ $(<n)$, *then* $\hat{\sigma}^2$ *may be also rewritten as* $\hat{\sigma}^2 = \mathbf{Y}'(\mathbf{I}_n - \mathbf{X}(\mathbf{X}'\mathbf{X})^{-1}\mathbf{X}')\mathbf{Y}/(n - p)$, *where* $\mathbf{I}_n = [\delta_{ij}]_{n\times n}$, $\delta_{ii} = 1$, $\delta_{ij} = 0$, *for* $i \neq j, \ldots$.

Theorem (Under Normality Assumption). *If in addition to the stated conditions in the Gauss–Markov theorem, the* ε_i *are normally distributed, and rank* $\mathbf{X} = p$, *then* $\hat{\beta}_i$ *has minimum variance in the class of all unbiased* (*linear or not*) *estimators of* β_i, *for* $i = 1, \ldots, p$. *Also if* $p < n$, *then* $\hat{\sigma}^2 = (\mathbf{Y} - \mathbf{X}\hat{\boldsymbol{\beta}})'(\mathbf{Y} - \mathbf{X}\hat{\boldsymbol{\beta}})/(n - p)$ *is a minimum-variance unbiased estimator of* σ^2. *Finally,* $\hat{\boldsymbol{\beta}} = (\hat{\beta}_1, \ldots, \hat{\beta}_p)'$ *has a* $N(\boldsymbol{\beta}, \sigma^2(\mathbf{X}'\mathbf{X})^{-1})$ *distribution, the* $\hat{\beta}_i$ *are independent of* $\hat{\sigma}^2 = (\mathbf{Y} - \mathbf{X}\hat{\boldsymbol{\beta}})'(\mathbf{Y} - \mathbf{X}\hat{\boldsymbol{\beta}})/(n - p)$, *and* $(n - p)\hat{\sigma}^2/\sigma^2$ *has a chi-square distribution of* $(n - p)$ *degrees of freedom. The latter in particular means that* $(\hat{\beta}_i - \beta_i)/\hat{\sigma}^2([\mathbf{X}'\mathbf{X}]^{-1})_{ii}$ *has a Student distribution of* $(n - p)$ *degrees of freedom, and confidence intervals for* β_i *may be set up.*

Least-squares Estimates in the Presence of Linear Constraints. The following situation often arises in the linear model $\mathbf{Y} = \mathbf{X}\boldsymbol{\beta} + \boldsymbol{\varepsilon}$, where one is interested in finding a least-squares estimate of $\boldsymbol{\beta}$ under a linear constraint of the form $\mathbf{C}\boldsymbol{\beta} = 0$, where $\mathbf{C}$ is an $s \times p$ $(s \leqslant p)$ matrix with rank $(\mathbf{C}) = s$. In this case one may use the method of Lagrange multipliers $\lambda_1, \ldots, \lambda_s$ to find a least-squares estimate of $\boldsymbol{\beta}$. This leads to modified normal equations: $\mathbf{X}'\mathbf{X}\boldsymbol{\beta} + \mathbf{C}'\boldsymbol{\lambda} = \mathbf{X}'\mathbf{Y}$, $\mathbf{C}\boldsymbol{\beta} = 0$, where $\boldsymbol{\lambda} = (\lambda_1, \ldots, \lambda_s)'$. It is in general easier to treat each such a problem directly and separately, than to consider a general formulation.

Canonical Form of the Linear Model. Let

$$\mathbf{Y} = \boldsymbol{\theta} + \boldsymbol{\varepsilon}, \qquad \boldsymbol{\theta} = \mathbf{X}\boldsymbol{\beta}, \qquad E[\varepsilon_i] = 0,$$

$$\sigma^2(\varepsilon_i) = \sigma^2 < \infty, \quad i = 1, \ldots, n, \qquad \mathrm{Cov}[\varepsilon_i, \varepsilon_j] = 0 \quad \text{for } i \neq j.$$

Let $\mathbf{v}_1, \ldots, \mathbf{v}_r$ be an orthonormal basis for Ω_r, and $\mathbf{v}_1, \ldots, \mathbf{v}_r, \mathbf{v}_{r+1}, \ldots, \mathbf{v}_n$ be an orthonormal basis for V_n. Then we may write

$$\boldsymbol{\theta} = \sum_{i=1}^{r} a_i \mathbf{v}_i, \qquad \mathbf{Y} = \sum_{i=1}^{n} Z_i \mathbf{v}_i, \qquad Z_i = \mathbf{v}_i'\mathbf{Y}.$$

If we define $a_{r+1} = \cdots = a_n \equiv 0$, then $E[Z_i] = \mathbf{v}_i'\boldsymbol{\theta} = a_i$, $i = 1, \ldots, n$, $\sigma^2(Z_i) = \sigma^2$, $\mathrm{Cov}[Z_i, Z_j] = 0$. We may also write the linear model in the canonical form: $Z_i = a_i + \mathbf{v}_i'\boldsymbol{\varepsilon}$, $i = 1, \ldots, n$. If rank $\mathbf{X} = p$, then we note that $\hat{\boldsymbol{\beta}} = \sum_{i=1}^{p} (\mathbf{X}'\mathbf{X})^{-1}\mathbf{X}'\mathbf{v}_i Z_i$, and hence $\hat{\boldsymbol{\beta}}$ depends only on $Z_1, \ldots, Z_p$ and not on $Z_{p+1}, \ldots, Z_n$. Also

$$(\mathbf{Y} - \mathbf{X}\hat{\boldsymbol{\beta}}) = \sum_{i=p+1}^{n} Z_i \mathbf{v}_i,$$

and with $p < n$, $\hat{\sigma}^2 = \sum_{i=p+1}^{n} Z_i^2/(n-p)$, that is, $\hat{\sigma}^2$ depends only on $Z_{p+1}, \ldots, Z_n$ and not on $Z_1, \ldots, Z_p$. If the ε_i are normally distributed, that is, $\mathbf{Y}$ has a $N(\boldsymbol{\theta}, \sigma^2\mathbf{I}_n)$ distribution, then the Z_i are independent and Z_i has a $N(a_i, \sigma^2)$, for $i = 1, \ldots, n$, distribution, with $a_{r+1} = \cdots = a_n$, $r \leqslant p$. Also $(Z_1, \ldots, Z_r, \sum_{i=1}^{n} Z_i^2)'$ is a complete and sufficient statistic for $(a_1, \ldots, a_r, \sigma^2)'$, $r \leqslant p < n$.

Tests of Hypotheses. Suppose we are interested in testing the hypothesis that $\boldsymbol{\theta}$ lies in some subspace ω_s of Ω_r, where ω_s is of s-dimensions $s < r < n$.

To carry out such a test we suppose the ε_i in the linear model:

$$\mathbf{Y} = \boldsymbol{\theta} + \boldsymbol{\varepsilon}, \qquad \boldsymbol{\theta} = \mathbf{X}\boldsymbol{\beta},$$

$$\sigma^2(\varepsilon_i) = \sigma^2 < \infty, \quad i = 1, \ldots, n, \qquad \mathrm{Cov}[\varepsilon_i, \varepsilon_j] = 0 \quad \text{for } i \neq j,$$

are normally distributed. Let $\mathbf{v}_1, \ldots, \mathbf{v}_s$ be an orthonormal basis for ω_s, $\mathbf{v}_1, \ldots, \mathbf{v}_s, \mathbf{v}_{s+1}, \ldots, \mathbf{v}_r$ be an orthonormal basis for Ω_r, $\mathbf{v}_1, \ldots, \mathbf{v}_s, \mathbf{v}_{s+1}, \ldots, \mathbf{v}_r, \mathbf{v}_{r+1}, \ldots, \mathbf{v}_n$ be an orthonormal basis for V_n. In general, we may write $\boldsymbol{\theta} = \sum_{i=1}^{r} a_i \mathbf{v}_i \in \Omega_r$. Let

$$\mathbf{Y} = \sum_{i=1}^{n} \mathbf{v}_i Z_i, \qquad Z_i = \mathbf{v}_i'\mathbf{Y}, \quad \text{then} \quad E[Z_i] = a_i,$$

with $a_{r+1} = \cdots = a_n \equiv 0$. The hypothesis that $\boldsymbol{\theta}$ lies in ω_s is equivalent to the hypothesis that $a_{s+1} = \cdots = a_r = 0$. Let $\hat{\boldsymbol{\theta}}_0$ minimize the expression $\|\mathbf{Y} - \boldsymbol{\theta}\|^2 \equiv \sum_{i=1}^n (Y_i - \theta_i)^2$ with $\boldsymbol{\theta} \in \omega_s$, and $\hat{\boldsymbol{\theta}}_1$ minimize the expression $\|\mathbf{Y} - \boldsymbol{\theta}\|^2 \equiv \sum_{i=1}^n (Y_i - \theta_i)^2$ with $\boldsymbol{\theta} \in \Omega_r$. Then the likelihood ratio test (§1.35) leads to the statistic:

$$F = \left(\frac{n - r}{r - s}\right) \frac{(\|\mathbf{Y} - \hat{\theta}_0\|^2 - \|\mathbf{Y} - \hat{\boldsymbol{\theta}}_1\|^2)}{\|\mathbf{Y} - \hat{\boldsymbol{\theta}}_1\|^2}$$

and rejects the hypothesis for large values of this statistic. In terms of the canonical variables $Z_1, \ldots, Z_n$, it may be rewritten as, with $r < n$,

$$F = \frac{\left[\sum_{i=s+1}^{r} Z_i^2/(r - s)\right]}{\left[\sum_{i=r+1}^{n} Z_i^2/(n - r)\right]}.$$

From the property that the Z_i are independent normally distributed $N(a_i, \sigma^2)$, with $a_{r+1} = \cdots = a_n \equiv 0$, we conclude that F has a noncentral Fisher's F-distribution with $(r - s)$ and $(n - r)$ degrees of freedom and a noncentrality parameter $\delta = \sum_{i=s+1}^r a_i^2/\sigma^2$. The noncentrality parameter δ may be also rewritten as $\delta = \|\boldsymbol{\theta} - \boldsymbol{\theta}_0\|^2/\sigma^2$, where $\boldsymbol{\theta}_0$ is the projection of $\boldsymbol{\theta}$ on ω_s, that is, if we write $\boldsymbol{\theta} = \sum_{i=1}^r \mathbf{v}_i a_i$, then $\boldsymbol{\theta}_0 = \sum_{i=1}^s \mathbf{v}_i a_i$. If the hypothesis $\boldsymbol{\theta} \in \omega_s$ is true, then $a_{s+1} = \cdots = a_r = 0$, and $\delta = 0$. The power of the test is an increasing function of δ and hence the test is unbiased; also the test is consistent.

(i) *Regression Analysis.* The linear model $\mathbf{Y} = \mathbf{X}\boldsymbol{\beta} + \boldsymbol{\varepsilon}$ may be written in terms of the components of $\mathbf{Y}$ as $Y_i = X_{i1}\beta_1 + \cdots + X_{ip}\beta_p + \varepsilon_i$, $i = 1, \ldots, n$. In regression models, it is assumed that the X_{ij} are variables that may be controlled and they are called the regressors, and the Y_i as the response variables. The matrix $\mathbf{X}$ is referred to as the regression matrix. A regression model is a linear model in which the random variables Y_i depend functionally on *quantitative* variables (the regressors). For example, Y_i may stand for a measurement made at a given (controllable) temperature T_i, and Y_i may have a functional dependence on T_i itself, with $p = 3$, $X_{i1} = 1$, $X_{i2} = T_i$, $X_{i3} = (T_i)^2$, $Y_i = \beta_1 + \beta_2 T_i + \beta_3 (T_i)^2 + \varepsilon_i$. The latter linear (in $\beta_1, \beta_2, \beta_3$) model provides a quadratic regression model.

The simplest regression model is the straight-line regression one defined by: $Y_i = \beta_1 + \beta_2 x_i + \varepsilon_i$, $i = 1, \ldots, n$, where x_i is the regressor. By introducing the regression matrix

$$\mathbf{X} = \begin{bmatrix} 1 & x_1 \\ 1 & x_2 \\ \vdots & \vdots \\ 1 & x_n \end{bmatrix},$$

and setting $\boldsymbol{\beta} = (\beta_1, \beta_2)'$, we may write the straight-line regression model in the familiar form: $\mathbf{Y} = \mathbf{X}\boldsymbol{\beta} + \boldsymbol{\varepsilon}$. As before we assume that $E[\varepsilon_i] = 0$, $\sigma^2(\varepsilon_i) = \sigma^2 < \infty$, $\text{Cov}[\varepsilon_i, \varepsilon_j] = 0$ for $i \neq j$. If not all the $x_1, \ldots, x_n$ are equal, then rank $\mathbf{X} = 2$. The least-squares estimates of β_1 and β_2 are then respectively,

$$\hat{\beta}_1 = \bar{Y} - \hat{\beta}_2 \bar{x}, \qquad \hat{\beta}_2 = \frac{\sum_{i=1}^{n} (Y_i - \bar{Y})(x_i - \bar{x})}{\sum_{i=1}^{n} (x_i - \bar{x})^2}.$$

We wish to test the hypothesis H_0: $\beta_2 = 0$ against H_A: $\beta_2 \neq 0$ under the normality condition of the ε_i. The F-statistic is given by

$$F = \frac{(n-2) \sum_{i=1}^{n} (\hat{\beta}_1 + \hat{\beta}_2 x_i - \bar{Y})^2}{\sum_{i=1}^{n} (Y_i - \hat{\beta}_1 - \hat{\beta}_2 x_i)^2}$$

and has a noncentral F-distribution of 1 and $n - 2$ degrees of freedom, since $r = 2$, $s = 1$, with a noncentrality parameter $\delta = (\beta_2/\sigma)^2 \sum_{i=1}^{n} (x_i - \bar{x})^2$. By setting $\beta_2 = 0$, and hence $\delta = 0$, the null hypothesis H_0 is then rejected for large values of F.

(ii) *Analysis of Variance*. In an analysis of variance model, the observations Y_i have no functional dependence on independent variables. For example, in comparing the durability of k different tires (with mean durabilities $\mu_1, \ldots, \mu_k$) the observations are supposed not to have an explicit functional dependence on the weight of the vehicule used in the experiment. The matrix $\mathbf{X}$ in the linear model $\mathbf{Y} = \mathbf{X}\boldsymbol{\beta} + \boldsymbol{\varepsilon}$, is then referred to as the design matrix, and its matrix elements consist, generally, of 0's and 1's. The simplest analysis of variance model is of the form: $Y_{ij} = \mu_i + \varepsilon_{ij}$, where $j = 1, \ldots, n_i$, $i = 1, \ldots, k$. In a typical application, $\mu_1, \ldots, \mu_k$ correspond to the means of the distributions associated with k populations, and $n_1, \ldots, n_k$ denote the sample sizes selected from these populations, respectively.

One would then be interested in testing the hypothesis: H_0: $\mu_1 = \cdots = \mu_k$ ($= \mu$ unspecified) against H_A: at least two of the μ_i are not equal. This model is called the one-way analysis of variance model. It may be written in the familiar form: $\mathbf{Y} = \mathbf{X}\boldsymbol{\beta} + \boldsymbol{\varepsilon}$, where $\mathbf{Y} = (Y_{11}, \ldots, Y_{1n_1}, Y_{21}, \ldots, Y_{kn_k})'$, $\boldsymbol{\beta} = (\mu_1, \ldots, \mu_k)'$, $\boldsymbol{\varepsilon} = (\varepsilon_{11}, \ldots, \varepsilon_{1n_1}, \varepsilon_{21}, \ldots, \varepsilon_{kn_k})$, and the design matrix is given by

$$\mathbf{X} = \begin{bmatrix} \mathbf{1}_1 & \mathbf{0} & . & . & . & . & \mathbf{0} \\ \mathbf{0} & \mathbf{1}_2 & \mathbf{0} & . & . & . & \mathbf{0} \\ . & . & . & . & . & . & . \\ . & . & . & . & . & . & . \\ . & . & . & . & . & . & \mathbf{0} \\ \mathbf{0} & \mathbf{0} & \mathbf{0} & . & . & \mathbf{0} & \mathbf{1}_k \end{bmatrix}_{n \times k},$$

with $\mathbf{1}_i = (1, 1, \dots, 1)'$, containing n_i 1's, and $n = \sum_{i=1}^k n_i$. Clearly, rank $\mathbf{X} = k$. The least-squares estimates of $\mu_1, \dots, \mu_k$ are given by $\bar{Y}_{1.}, \dots, \bar{Y}_{k.}$, respectively, where $\bar{Y}_{i.} = \sum_{j=1}^{n_i} Y_{ij}/n_i$. Under H_0, the least-squares estimate of μ is $\bar{Y}_{..}$. Here $r = p = k$, $s = 1$, and the F-statistic is given by

$$F = \frac{\left[\left(\sum_{i=1}^k \frac{Y_{i.}^2}{n_i} - \frac{Y_{..}^2}{n}\right)\Big/(k-1)\right]}{\left[\left(\sum_{i=1}^k \sum_{j=1}^{n_i} Y_{ij}^2 - \sum_{i=1}^k \frac{Y_{i.}^2}{n_i}\right)\Big/(n-k)\right]}$$

and the noncentrality parameter is given by

$$\delta = \sum_{i=1}^k n_i(\mu_i - \bar{\mu})^2/\sigma^2,$$

where $Y_{i.} = \sum_{j=1}^{n_i} Y_{ij}$, $Y_{..} = \sum_{i=1}^k Y_{i.}$, $\bar{\mu} = \sum_{i=1}^k n_i\mu_i/n$. Also

$$\hat{\sigma}^2 = \frac{\left[\sum_{i=1}^k \sum_{j=1}^{n_i} Y_{ij}^2 - \sum_{i=1}^k (Y_{i.}^2/n_i)\right]}{(n-k)},$$

and hence $E[\hat{\sigma}^2] = \sigma^2$. Finally,

$$E\left[\sum_{i=1}^k \frac{Y_{i.}^2}{n_i} - \frac{Y_{..}^2}{n}\right] = \sum_{i=1}^k n_i(\mu_i - \bar{\mu})^2 + (k-1)\sigma^2.$$

A different parametrization of the one-way analysis of variance model is possibe. For we may write $\mu_i = \mu + (\mu_i - \mu)$, where $\mu \equiv \bar{\mu} = \sum_{i=1}^n n_i\mu_i/n$. Accordingly, we may define the model as $Y_{ij} = \mu + \alpha_i + \varepsilon_{ij}$ such that $\sum_{i=1}^k n_i\alpha_i = 0$. In this case, we may write $\boldsymbol{\beta} = (\mu, \alpha_1, \dots, \alpha_k)'$, and the design matrix $\mathbf{X}$ as

$$\mathbf{X} = \begin{bmatrix} \mathbf{1}_1 & \mathbf{1}_1 & \mathbf{0} & \mathbf{0} & . & . & . & . & \mathbf{0} \\ \mathbf{1}_2 & \mathbf{0} & \mathbf{1}_2 & \mathbf{0} & . & . & . & . & \mathbf{0} \\ . & . & . & . & . & . & . & . & . \\ . & . & . & . & . & . & . & . & . \\ . & . & . & . & . & . & . & . & \mathbf{0} \\ \mathbf{1}_k & \mathbf{0} & \mathbf{0} & \mathbf{0} & . & . & . & \mathbf{0} & \mathbf{1}_k \end{bmatrix}_{n\times(k+1)},$$

which clearly has rank $k < p \equiv (k+1)$. The least-squares estimates of μ, α_1, $\dots$, α_k, under the constraint $\mathbf{C\beta} = 0$, where $\mathbf{C} = (0, n_1, \dots, n_k)$, by the method of Lagrange multipliers, are obtained to be $\hat{\mu} = \bar{Y}_{..}$, $\hat{\alpha}_i = \bar{Y}_{i.} - \bar{Y}_{..}$. The test then becomes H_0: $\alpha_1 = \cdots = \alpha_k = 0$. More complicated analysis of variance models are similarly handled [cf. Fisher and McDonald (1978), Scheffé (1959)].

(iii) *Analysis of Covariance.* In analysis of covariance models, the Y's have dependence on qualitative and quantitative variables. For example, in comparing the lifetime of k different tires (qualitative variable), the observations themselves may have an explicit functional dependence on the weight (quantitative variable) of the vehicle used in the experiment. As a simple covariance model, we consider $Y_{ij} = \mu_i + \eta x_{ij} + \varepsilon_{ij}, j = 1, \ldots, n_i, i = 1, \ldots, k$, where for at least one i, not all the $x_{i1}, \ldots, x_{in_i}$ are equal. We may write $\mathbf{Y} = (Y_{11}, \ldots, Y_{1n_1}, Y_{21}, \ldots, Y_{kn_k})'$, $\boldsymbol{\varepsilon} = (\varepsilon_{11}, \ldots, \varepsilon_{1n_1}, \varepsilon_{21}, \ldots, \varepsilon_{kn_k})'$, $\boldsymbol{\beta} = (\mu_1, \ldots, \mu_k, \eta)'$,

$$\mathbf{X} = \begin{bmatrix} \mathbf{1}_1 & \mathbf{0} & . & . & . & \mathbf{0} & \mathbf{X}_1 \\ \mathbf{0} & \mathbf{1}_2 & \mathbf{0} & . & . & \mathbf{0} & \mathbf{X}_2 \\ . & \mathbf{0} & . & . & . & . & . \\ . & . & . & . & . & . & . \\ . & . & . & . & . & \mathbf{0} & . \\ \mathbf{0} & \mathbf{0} & . & . & . & \mathbf{1}_k & \mathbf{X}_k \end{bmatrix}_{n \times (k+1)},$$

where $\mathbf{1}_i = (1, \ldots, 1)'$ with n_i 1's, $\mathbf{X}_i = (x_{i1}, \ldots, x_{in_i})'$. We may then write the model in the familiar form: $\mathbf{Y} = \mathbf{X}\boldsymbol{\beta} + \boldsymbol{\varepsilon}$. Clearly $\operatorname{rank}(\mathbf{X}) = k + 1$. The least-squares estimates of η and μ_i are given respectively, by $\hat{\eta} = N_{xY}/N_{xx}$, where

$$N_{xx} = \sum_{i=1}^{k} \sum_{j=1}^{n_i} x_{ij}^2 - \sum_{i=1}^{k} (x_{i.}^2/n_i),$$

$$N_{xY} = \sum_{i=1}^{k} \sum_{j=1}^{n_i} x_{ij} Y_{ij} - \sum_{i=1}^{k} (x_{i.} Y_{i.}/n_i),$$

and

$$\hat{\mu}_i = \bar{Y}_{i.} - \bar{x}_{i.}(N_{xY}/N_{xx}).$$

We wish to test the hypothesis H_0: $\mu_1 = \cdots = \mu_k$ ($= \mu$ unspecified) against H_A: not all the μ_i are equal. The F-statistic then is given by

$$F = \frac{N/(k-1)}{D/(n-k-1)},$$

$$N = \left(\sum_{i=1}^{k} \frac{Y_{i.}^2}{n_i} - \frac{Y_{..}^2}{n} \right) - \frac{\left(\sum_{i=1}^{k} \sum_{j=1}^{n_i} x_{ij} Y_{ij} - \frac{x_{..} Y_{..}}{n} \right)^2}{\left(\sum_{i=1}^{k} \sum_{j=1}^{n_i} x_{ij}^2 - \frac{x_{..}^2}{n} \right)} + \frac{N_{xY}^2}{N_{xx}},$$

$$D = \sum_{i=1}^{k} \sum_{j=1}^{n_i} Y_{ij}^2 - \frac{Y_{..}^2}{n} - \frac{N_{xY}^2}{N_{xx}}.$$

The degrees of freedom $(k - 1)$, $(n - k - 1)$ are readily obtained by noting that $r = p = (k + 1)$, $s = 2$. We also note that $D/(n - k - 1) = \hat{\sigma}^2$.

(iv) *Robustness of the F-test*. The F-test is, in general, robust to departures from the normality assumption of the errors ε_i. [Also in the k-sample Pitman randomization test (§§1.43 and 3.13), the F'-statistic, based on no assumption on the distribution of the underlying population, has, for $n_1, \ldots, n_k$ large, approximately the first two moments of the F-statistic in the one-way analysis of variance model of $(k - 1)$ and $(n - k)$ degrees of freedom for $n = n_1 +$

$\cdots + n_k$ large.] At least for the one-way analysis of variance, the F-test is robust to departures of the equality of variances of the ε_i as long as the sample sizes $n_1, \ldots, n_k$ are equal. The test is however, in general, sensitive to departures of the equality of variances if the sample sizes are unequal. [For the test of the equality of variances one may use the robust jackknife method (see §1.42(iii)).] The F-test, however, seems, in general, sensitive to departures from the assumption of the absence of correlations between the ε_i, and the presence of such correlations may have large effect on conclusions based on the F-test. For detailed treatments of the robustness of the F-test see, e.g., Seber (1977) Scheffé (1959).

(v) *Simultaneous Confidence Intervals.*

Bonferroni's Inequality. Let I_i be a $100(1 - \alpha_i)\%$ confidence interval for the parameter β_i in the linear model $\mathbf{Y} = \mathbf{X\beta} + \varepsilon$. We have already discussed in the case when rank $\mathbf{X} = p$ and under the normality condition how such a confidence interval for β_i may be set based on the least-squares estimate $\hat{\beta}_i$ of β_i. We may write $P[\{\beta_i \in I_i\}] \geqslant 1 - \alpha_i$ for all $\mathbf{\beta} \in \Omega_p$. Then we have the inequality

$$P\left[\bigcap_{i=1}^{p} \{\beta_i \in I_i\}\right] \geqslant 1 - \sum_{i=1}^{k} \alpha_i.$$

That is, if we set $\alpha = \sum_{i=1}^{p} \alpha_i$, then with a confidence coefficient at least equal to $1 - \alpha$, the intervals $I_1, \ldots, I_p$ provide simultaneous confidence intervals for $\beta_1, \ldots, \beta_p$. These intervals, referred to as Bonferroni intervals, however are, in general, too wide and better techniques may be developed as given by Scheffé's and Tukey's methods.

Scheffé's Method. In the linear model $\mathbf{Y} = \mathbf{X\beta} + \varepsilon$, suppose rank $\mathbf{X} = p < n$, then under the normality condition we have for any linear parametric function $\mathbf{w}'\mathbf{\beta}$, that is for any (non-zero) p-vector $\mathbf{w}$ in Ω_p, the Scheffé's confidence interval with confidence coefficient $1 - \alpha$ (obtained by an application, in the process, of the Cauchy–Schwarz inequality) given by

$$(\mathbf{w}'\hat{\mathbf{\beta}} \pm \sqrt{p\hat{\sigma}^2 F_\alpha(p, n-p)(\mathbf{w}'(\mathbf{X}'\mathbf{X})^{-1}\mathbf{w})}),$$

where $F_\alpha(p, n-p)$ is the $(1-\alpha)$th quantile of the F-distribution of p and $n-p$ degrees of freedom: $P[F \geqslant F_\alpha(p, n-p)] = \alpha$.

Tukey's Method. Consider the one-way analysis of variance model, under the normality assumption, $Y_{ij} = \mu_i + \varepsilon_{ij}$, with $n_1 = n_2 = \cdots = n_k = m$. If the hypothesis H_0: $\mu_1 = \cdots = \mu_k$ is rejected, then it is interesting to compare the μ_i's pairwise. By an application of the Cauchy–Schwarz inequality one arrives at the conclusion that the $k(k-1)/2$ inequalities

$$|\bar{Y}_{i.} - \bar{Y}_{j.} - (\mu_i - \mu_j)| \leqslant \sqrt{\frac{\hat{\sigma}^2}{m}}\, q_{k,k(m-1)}(1-\alpha) \qquad (*)$$

hold simultaneously with a probability not smaller than $1 - \alpha$, where $q_{k,k(m-1)}(\alpha)$ is the $(1 - \alpha)$th quantile of a Studenized range variable (§3.45) of $k(m - 1)$ degrees of freedom and parameter k ($\equiv n$ in §3.45). By definition $P[U \geqslant q_{k,k(m-1)}(\alpha)] = \alpha$.

If for any pair (i, j), $i \neq j$, in $(1, \ldots, k)$, we set $\mu_i = \mu_j$ and the inequality in $(*)$ is violated we infer that $\mu_i \neq \mu_j$. The inequalities in $(*)$ provide simultaneous confidence intervals, with confidence coefficient not less than $1 - \alpha$, for the differences $(\mu_i - \mu_j)$:

$$\left(\bar{Y}_{i.} - \bar{Y}_{j.} \pm \sqrt{\frac{\hat{\sigma}^2}{m}}\, q_{k,k(m-1)}(\alpha)\right).$$

One disadvantage of the Tukey method is that it requires that all the $n_1, \ldots, n_k$ are equal. Scheffé's method is more general, and is, in general, more robust to departures from the normality assumption and the equality of variances of the ε_i than the Tukey method. For the pairwise comparison problem $\mu_i = \mu_j$ the Tukey method provides, in general, shorter intervals than the Scheffé's method, as the latter is more general.

See also §1.44.

[Cf., Seber (1977, 1980), Scheffé (1959), Fisher and McDonald (1978), Fisher (1951), Roussas (1973), Bickel and Doksum (1977), Kendall and Stuart (1979), Silvey (1975), Miller (1981).]

§1.41. The Bootstrap and the Jackknife

Let $X_1, \ldots, X_n$ be independent identically distributed random variables each with a distribution F depending on a parameter θ. Let $\hat{G}(X_1, \ldots, X_n)$, symmetric in $X_1, \ldots, X_n$, be an estimator of $G(\theta)$, for some function G of θ.

We are interested in estimating expressions of the form $E_F[\hat{G}]$, $\sigma_F^2(\hat{G})$, bias $= E_F[\hat{G}] - G(\theta)$, $P_F[\hat{G} \leqslant C]$, etc.... We denote any such expression by $\lambda_F(\hat{G})$. Let $X_1 = x_1, \ldots, X_n = x_n$ denote the values obtained by a sample. Let $X_1^*, \ldots, X_n^*$ be a random sample from $[x_1, \ldots, x_n]$ *with* replacement, where the probability of choosing any x_i is a constant. That is,

$$P[X_1^* = a_1, \ldots, X_n^* = a_n] = \binom{n}{c_1 \ldots c_n} n^{-n},$$

where $a_1, \ldots, a_n$ are numbers, not necessarily distinct, with values from $[x_1, \ldots, x_n]$, such that c_1 of them equal to $x_1, \ldots, c_n$ of them equal to x_n, $c_1 + \cdots + c_n = n$. We note from the multinomial distribution, or directly, that

$$\sum{}^* c_i \binom{n}{c_1 \ldots c_n} n^{-n} = 1, \qquad \sum{}^* (c_i - 1)(c_j - 1) \binom{n}{c_1 \ldots c_n} n^{-n} = \left(\delta_{ij} - \frac{1}{n}\right),$$

where $\delta_{ii} = 1$, $\delta_{ij} = 0$ for $i \neq j$, and $\sum^*$ denotes a sum over all nonnegative integers $c_1, \ldots, c_n$ such that $c_1 + \cdots + c_n = n$. $X_1^*, \ldots, X_n^*$ is called a bootstrap sample.

Define $\hat{G}^* = \hat{G}(X_1^*, \ldots, X_n^*)$. The *bootstrap* estimate of $\lambda_F(\hat{G})$ is defined by $\lambda_*(\hat{G}^*)$ evaluated with respect to the probability $P[X_1^* = a_1, \ldots, X_n^* = a_n]$. That is, $E_F[G]$ is estimated by

$$E_*[\hat{G}^*] = \sum{}^* \hat{G}(a_1, \ldots, a_n)\binom{n}{c_1 \ldots c_n} n^{-n}.$$

For example, if $\hat{G}(X_1, \ldots, X_n) = \sum_{i=1}^n X_i/n = \bar{X}$, then $E_F[\bar{X}] = \mu$ is estimated by

$$E_*[\bar{X}^*] = \sum{}^* \frac{(c_1 x_1 + \cdots + c_n x_n)}{n}\binom{n}{c_1 \ldots c_n} n^{-n} = \bar{x},$$

where $\bar{X}^* = \sum_{i=1}^n X_i^*/n$. Similarly, $\sigma^2(\bar{X})$ is estimated by

$$\sigma_*^2(\bar{X}^*) = \sum{}^* \frac{(c_1 x_1 + \cdots + c_n x_n - x_1 - \cdots - x_n)^2}{n^2}\binom{n}{c_1 \ldots c_n} n^{-n}$$

$$= \sum_{i=1}^n (x_i - \bar{x})^2/n^2.$$

Also $P_F[\hat{G} \leqslant C] = E_F[I(\hat{G} \leqslant C)]$, where I is the indicator function $I(\hat{G}(a_1, \ldots, a_n) \leqslant C) = 1$ if $\hat{G}(a_1, \ldots, a_n) \leqslant C$ and $= 0$ otherwise. Hence $P_F[\hat{G} \leqslant C]$ is estimated by

$$P_*[\hat{G}^* \leqslant C] = \sum{}^* I(\hat{G}(a_1, \ldots, a_n) \leqslant C)\binom{n}{c_1 \ldots c_n} n^{-n}.$$

The bootstrap estimate of the bias $E_F[\hat{G}] - G(\theta)$ is given by $E_*[\hat{G}^*] - \hat{G}$.

We note that $\lambda(\hat{G})$ cannot always be written in a manageable form and approximations for $\lambda_*(\hat{G}^*)$ may be sought. Consider a large number B of replicates $\hat{G}_1^*, \ldots, \hat{G}_B^*$ based on B bootstrap samples chosen (with replacement) from $[x_1, \ldots, x_n]$. Then approximate expressions for $E_*[\hat{G}^*]$, $\sigma_*^2(\hat{G}^*)$, $E_*[\hat{G}^*] - \hat{G}$, $P_*[\hat{G}^* \leqslant C] = E_*[I(\hat{G}^* \leqslant C)]$, respectively, are:

$$\sum_{i=1}^B \hat{G}_i^*/B \equiv \hat{G}_\cdot^*, \qquad \sum_{i=1}^B (\hat{G}_i^* - \hat{G}_\cdot^*)^2/B,$$

$$\hat{G}_\cdot^* - \hat{G}, \qquad \sum_{i=1}^B I(\hat{G}_i^* \leqslant C)/B = [\text{number of } \hat{G}_i^* \leqslant C]/B.$$

Unfortunately, the bootstrap method involves many computations. This brings us to the jackknife method which requires less computations.

The Jackknife Method

Any vector $\mathbf{P}^* = (P_1^*, \ldots, P_n^*)'$ such that $P_i^* \geqslant 0$, $i = 1, \ldots, n$ and $P_1^* + \cdots + P_n^* = 1$ is called a *resampling vector*. We are interested in three types of resampling vectors: $\mathbf{P}_B^* = (c_1/n, \ldots, c_n/n)'$, where $c_1, \ldots, c_n$ are nonnegative integers such that $c_1 + \cdots + c_n = n$; $\mathbf{P}_0 = (1/n, \ldots, 1/n)'$ (which is a special case of $\mathbf{P}_B^*$ but will be singled out for convenience);

$$\mathbf{P}_{(i)} = \left(\frac{1}{n-1}, \ldots, \frac{1}{n-1}, 0, \frac{1}{n-1}, \ldots, \frac{1}{n-1}\right)'$$

with 0 at the ith place.

We may then conveniently write

$$\hat{G}(X_1^*, \ldots, X_n^*) = \hat{G}(\mathbf{P}_B^*), \qquad \hat{G}(X_1, \ldots, X_n) = \hat{G}(P_0).$$

Similarly, let $\hat{G}_{(i)}$ denote the statistic based on the sample $X_1, \ldots, X_{i-1}, X_{i+1}, \ldots, X_n$ of size $(n-1)$ by deleting the ith observation.

For example, if $\hat{G} = \sum_{j=1}^n X_j/n$, then

$$\hat{G}_{(i)} = \sum_{\substack{j=1 \\ j \neq i}}^{n} X_j/(n-1) \equiv \bar{X}_i,$$

and if $\hat{G} = \sum_{j=1}^n (X_j - \bar{X})^2/(n-1)$, then

$$\hat{G}_{(i)} = \sum_{\substack{j=1 \\ j \neq i}}^{n} (X_j - \bar{X}_i)^2/(n-2) \equiv S_{-i}^2.$$

We write $\hat{G}_{(i)} = \hat{G}(\mathbf{P}_{(i)})$.

Suppose we may approximate $\hat{G}(\mathbf{P}_B^*)$ by a quadratic form:

$$\hat{G}_Q(\mathbf{P}^*) = \hat{G}(\mathbf{P}_0) + (\mathbf{P}^* - \mathbf{P}_0)'\mathbf{U} + \tfrac{1}{2}(\mathbf{P}^* - \mathbf{P}_0)'\mathbf{V}(\mathbf{P}^* - \mathbf{P}_0),$$

for $\mathbf{P}^* = \mathbf{P}_B^*$, where $\mathbf{U} = (\mathbf{U}_1, \ldots, U_n)'$ is an n-vector and $\mathbf{V} = [V_{ij}]$ is a symmetric $(n \times n)$-matrix, *such that* $\hat{G}_Q(\mathbf{P}_{(i)}) = \hat{G}(\mathbf{P}_{(i)})$ for all $i = 1, \ldots, n$. We note that $\hat{G}_Q(\mathbf{P}_0) = \hat{G}(\mathbf{P}_0)$. Since $\sum_{i=1}^n (c_i - 1) = 0$, we may without loss of generality suppose that

$$\sum_{i=1}^{n} U_i = 0, \qquad \sum_{i=1}^{n} V_{ij} = 0, \qquad \sum_{j=1}^{n} V_{ij} = 0.$$

Since $\hat{G}_Q(\mathbf{P}^*)$ is quadratic in $\mathbf{P}^*$ we may readily evaluate $E_*[\hat{G}_Q(\mathbf{P}_B^*)]$ (in contradistinction with a more general expression $E_*[\hat{G}(\mathbf{P}_B^*)]$).

Theorem (Efron). *Since* $\hat{G}_Q(\mathbf{P}_{(i)} = \hat{G}(\mathbf{P}_{(i)})$ *we have explicitly,*

$$E_*[\hat{G}_Q(\mathbf{P}_B^*)] = \hat{G}(\mathbf{P}_0) + \frac{(n-1)^2}{n}\left[\sum_{i=1}^{n} \frac{\hat{G}(\mathbf{P}_{(i)})}{n} - \hat{G}(\mathbf{P}_0)\right].$$

Hence we have for the "bias"

$$E_*[\hat{G}_Q(\mathbf{P}_B^*)] - \hat{G}(\mathbf{P}_0) = \left(1 - \frac{1}{n}\right)(n-1)\left[\sum_{i=1}^{n} \frac{\hat{G}(\mathbf{P}_{(i)})}{n} - \hat{G}(\mathbf{P}_0)\right].$$

We note that for n sufficiently large $(1 - 1/n) \simeq 1$, and the expression for

$E_*[\hat{G}_Q(\mathbf{P}_B^*)]$ is "approximately" equal to

$$(n-1)\left[\sum_{i=1}^{n} \frac{\hat{G}(\mathbf{P}_{(i)})}{n} - \hat{G}(\mathbf{P}_0)\right].$$

The latter expression is called the *Quenouille estimate of the bias*. The *jackknife estimate* of $G(\theta)$ is *defined* by subtracting the Quenouille bias from $\hat{G}(\mathbf{P}_0)$, that is, it is defined by:

$$\tilde{G}_. = \hat{G}(P_0) - (n-1)\left[\sum_{i=1}^{n} \frac{\hat{G}(\mathbf{P}_{(i)})}{n} - \hat{G}(\mathbf{P}_0)\right]$$

or

$$\tilde{G}_. = n\hat{G}(\mathbf{P}_0) - (n-1)\sum_{i=1}^{n} \frac{\hat{G}(\mathbf{P}_{(i)})}{n}.$$

This suggests to define

$$\tilde{G}_i = n\hat{G}(\mathbf{P}_0) - (n-1)\hat{G}(\mathbf{P}_{(i)}),$$

and write

$$\tilde{G}_. = \sum_{i=1}^{n} \tilde{G}_i/n$$

for the jackknife statistic. $\tilde{G}_1, \ldots, \tilde{G}_n$ are called the jackknife pseudo-values. We note that we may rewrite $\tilde{G}_i$ as

$$\tilde{G}_i = n\hat{G}_n(X_1, \ldots, X_n) - (n-1)\hat{G}_{n-1}(X_1, \ldots, X_i, X_{i+1}, \ldots, X_n),$$

where we have written subscripts n and $n-1$ in $\hat{G}$ to emphasize the corresponding sample sizes. Although the jackknife estimate $\tilde{G}_.$ of $G(\theta)$ was obtained by a quadratic approximation in $\mathbf{P}_B^*$ of the bootstrap estimation method of $G(\theta)$, the jackknife statistic $\tilde{G}_.$ may be defined independently and is an estimator of $G(\theta)$ in its own right. Clearly the jackknife method requires less computations than the bootstrap method.

As an example, let $\hat{G}_n(X_1, \ldots, X_n) = f(\bar{X})$, then

$$\hat{G}_i = nf(\bar{X}) - (n-1)f(\bar{X}_i), \qquad \bar{X}_i = \sum_{\substack{j=1\\ j\neq i}}^{n} X_j/(n-1).$$

The jackknife statistic was initially introduced to reduce bias. For example, if

$$E_F[\hat{G}_n] = G + \frac{a_1(F)}{n} + \frac{a_2(F)}{n^2} + O(n^{-2-\delta})$$

for some $1 \geqslant \delta > 0$, where $a_1(F)$ and $a_2(F)$ are independent of n. Then

$$E_F[\hat{G}_{n-1}] = G + \frac{a_1(F)}{n-1} + \frac{a_2(F)}{(n-1)^2} + O(n^{-2-\delta}),$$

and hence

$$E_F[\tilde{G}_.] = G + O(n^{-1-\delta}),$$

thus reducing the bias from $O(n^{-1})$ to $O(n^{-1-\delta})$. [As an illustration of this property suppose that $\hat{G}_n = \sum_{i=1}^n (X_i - \bar{X})^2/n$. Then $E_F[\hat{G}_n] = \sigma^2 - \sigma^2/n$, and hence $a_1(F) = -\sigma^2$, where $\sigma^2(X_i) \equiv \sigma^2$,

$$\sum_{i=1}^n \hat{G}_{n-1}(x_1, \ldots, x_{i-1}, x_{i+1}, \ldots, x_n)/n = \frac{n(n-2)}{(n-1)^2} \hat{G}_n,$$

and

$$\tilde{G}_. = \sum_{i=1}^n (X_i - \bar{X})^2/(n-1)$$

giving the familiar unbiased estimator of σ^2.] The jackknife estimate of $\sigma_F^2(\hat{G})$ is defined by $\sum_{i=1}^n (\tilde{G}_i - \tilde{G}_.)^2/(n-1)$.

The jackknife estimator, in general, is not as efficient (Efron) as the bootstrap estimate. For asymptotic theorems of jackknife statistics see §2.33.

Finally we note that if we have observations

$$X_1 = x_1, \ldots, X_{i-1} = x_{i-1}, \qquad X_{i+1} = x_{i+1}, \ldots, X_n = x_n,$$

then for n not too small we may write $\tilde{G}_i(x_1, \ldots, x_n) \approx \mathrm{SC}(x)$, where

$$\mathrm{SC}(x) = n[\hat{G}_n(x_1, \ldots, x_{i-1}, x, x_{i+1}, \ldots, x_n) - \hat{G}_{n-1}(x_1, \ldots, x_{i-1}, x_{i+1}, \ldots, x_n)] \quad \text{for } x_i = x,$$

which measures the change in the value taken by the statistic $\hat{G}_{n-1}$ when an additional observation $x_i = x$ is included in the sample. SC(x) is called the sensitivity curve of the estimator $\hat{G}_n$ (see also §1.42).

In particular, if $\hat{G}_n = \sum_{j=1}^n x_j/n$, then

$$\mathrm{SC}(x) = \left(x - \frac{x_1 + \cdots + x_{i-1} + x_{i+1} + \cdots + x_n}{n-1}\right)$$

and a large value of x may lead to a large value for SC(x) indicating the "sensitivity" of the mean $\bar{X}$ to "outliers."

See also §§1.12, 1.42, 2.19, 2.25, and 2.33.

[Cf., Efron (1979, 1982, 1983), Gray and Schucany (1972), Hampel (1968), Miller (1964, 1968, 1974), Serfling (1980), Quenouille (1956), Manoukian (1986).]

§1.42. Robustness

A statistical procedure is said to be *robust*, with respect to a given assumption, if the procedure is not "too sensitive" to departures from its given assumption. For example, for sufficiently large samples, and certain regularity conditions (§§3.47, 3.48) the distribution of the T-statistic is "fairly robust" to departures from the normality assumption of the underlying population(s). [That is, for sufficiently large samples, one may still use critical values from tables of the standard normal distribution.] On the other hand, the classic (parametric) tests on variances of chi-square (based on $(n-1)S^2/\sigma_0^2$—see §3.48), F-test (based on S_1^2/S_2^2—see §3.48) and the Bartlett M-test (see §1.35), respectively, for the one-sample: $\sigma^2 = \sigma_0^2$, for the two-sample: $\sigma_1^2 = \sigma_2^2$, and for the k-sample: $\sigma_1^2 = \cdots = \sigma_k^2$, tests are quite sensitive [Box (1953), Miller (1968)] to departures from the normality assumption of the underlying populations. This result is already clear for the large-sample case from the asymptotic theorems (see §3.48[B]) for these statistics, which depend on the kurtosis γ_2 (assumed that $\gamma_2^{(1)} = \gamma_2^{(2)} = \cdots$) and modify the nominal critical values of tests if $\gamma_2 \neq 0$, which is indeed the case for nonnormal populations.

In general, a robust procedure (Huber) should have good efficiency and should yield for the latter a value near the optimal one when calculated at the model in question (usually the normal one). It should also not lead to drastic changes in the derived conclusions for large departures from the given underlying assumption.

(i) A good test of robustness, to departure from normality, for various statistical procedures may be carried out by examining Tukey's (1960a) contaminated normal distribution:

$$(1-\varepsilon)N(\mu, \sigma^2) + \varepsilon N(\mu_1, \sigma_1^2),$$

where $0 < \varepsilon < 1$. For $\mu = \mu_1$, the latter distribution is symmetric about μ, and if $\sigma^2 < \sigma_1^2$ ($\sigma^2 > \sigma_1^2$) the distribution has a heavier (lighter) tail than the corresponding one with $\varepsilon = 0$. And for a contaminated distribution, $(1-\varepsilon)N(\mu, \sigma^2) + \varepsilon N(\mu, \sigma_1^2)$, with $\sigma^2 < \sigma_1^2$, it takes a very small value $\varepsilon \neq 0$ so that the sample median becomes more efficient than the sample mean, with the latter as estimates of the location parameter. Explicitly, the asymptotic relative efficiency of the sample median to the sample mean is given by

$$\frac{2}{\pi}\left[1-\varepsilon+\varepsilon\left(\frac{\sigma_1}{\sigma}\right)^2\right]\left[1-\varepsilon+\varepsilon\left(\frac{\sigma_1}{\sigma}\right)\right]^2.$$

In §1.13, recommendations are made on how much trimming to be done in estimating a location parameter, by the α-trimmed mean, for a class of symmetric distributions including light-tailed (such as the normal) and heavy-tailed (such as the slash) distributions, with regard to efficiency, when the underlying distribution of the population is unknown, or when it is only known to be of light-tail or of a heavy-tail. (Here we recall that the sample mean and the sample median may be defined as special cases of α-trimmed means, see §1.13.)

(ii) *Sensitivity and Influence Curves.* Let $X_1, \ldots, X_n$ be independent and identically distributed random variables each with a distribution F. Let

$\hat{\theta}(X_1, \ldots, X_n)$ be an estimator of a certain parameter θ associated with the distribution F. Suppose $X_1 = x_1, \ldots, X_{n-1} = x_{n-1}$ are some observed fixed values, then we determine the "sensitivity" of the statistic $\hat{\theta}_n$ when an observation $X_n = x$ is included in the sample, by defining the change induced from $\hat{\theta}_{n-1}$ to $\hat{\theta}_n$ as

$$\mathrm{SC}(x) = n[\hat{\theta}_n(x_1, \ldots, x_{n-1}, x) - \hat{\theta}_{n-1}(x_1, \ldots, x_{n-1})]$$

as a function of x, where we have introduced the subscripts n and $n - 1$ to emphasize the corresponding sample sizes. SC(x) is called the *sensitivity curve* of the estimator $\hat{\theta}_n$.

For example, if θ is the mean and $\hat{\theta}_n = \bar{X}$ is the sample mean, then

$$\mathrm{SC}(x) = [x - (x_1 + \cdots + x_{n-1})/(n - 1)],$$

indicating the large effect on $\bar{X}$ from the addition of a relatively large observation (positive or negative) to the sample. The situation with the sample median as an estimate of a location parameter is quite different. Suppose, for example, that n is odd, then in the latter case:

$$\mathrm{SC}(x) = \begin{cases} \dfrac{n}{2}[y_{(n-1)/2} - y_{(n+1)/2}] & \text{if } x < y_{(n-1)/2}, \\ n\{x - \frac{1}{2}[y_{(n-1)/2} + y_{(n+1)/2}]\} & \text{if } y_{(n-1)/2} \leqslant x \leqslant y_{(n+1)/2}, \\ \dfrac{n}{2}[y_{(n+1)/2} - y_{(n-1)/2}] & \text{if } x > y_{(n+1)/2}, \end{cases}$$

where $y_1, y_2, \ldots, y_n$ are the ordered observations of $x_1, \ldots, x_n$: $y_1 \leqslant y_2 \leqslant \cdots \leqslant y_n$. This shows the less sensitivity of the sample median, as opposed to the sample mean, to an extreme observation.

Although SC(x) provides a summary of sensitivity or resistance of the statistic $\hat{\theta}_n$ to the addition of an observation, it depends on the sample and its size and hence is not always convenient. To obtain a more convenient measure of sensitivity or resistance not tied to a sample, directly from SC(x), we formally replace the empirical distribution by the distribution F and $\hat{\theta}_n$ by θ, assuming that $\hat{\theta}_n$ is a consistent estimator of θ. That is, formally we consider the limit $n \to \infty$. This leads to the concept of the *influence curve* (Hampel). To arrive to its definition, we rewrite $\hat{\theta}_{n-1}$ as $\hat{\theta}(\hat{F}_{n-1})$ where $\hat{F}_{n-1}$ is the empirical distribution which assigns mass $1/(n - 1)$ to each $x_1, \ldots, x_{n-1}$. If δ_x denotes the pointmass 1 at x, then we may write

$$\hat{F}_n = \frac{(n-1)}{n}\hat{F}_{n-1} + \frac{1}{n}\delta_x.$$

Thus we may rewrite SC(x) as:

$$\mathrm{SC}(x) = \frac{\hat{\theta}\left(\left(\dfrac{n-1}{n}\right)\hat{F}_{n-1} + \dfrac{1}{n}\delta_x\right) - \hat{\theta}(\hat{F}_{n-1})}{\dfrac{1}{n}}.$$

By replacing $1/n$ by ε ($n \to \infty$, $\varepsilon \to 0$), and $\hat{\theta}(\hat{F}_n)$ by $\theta(F)$ formally, where $\theta = \theta(F)$, we are led to the definition of the influence curve

$$\mathrm{IC}(x) = \lim_{\varepsilon \to 0} \left[\frac{\theta((1 - \varepsilon)F + \varepsilon\delta_x) - \theta(F)}{\varepsilon} \right].$$

$\mathrm{IC}(x)$ is a function of x and gives a measure of the influence of an observation on the value of an estimate without reference to a specific sample or sample size. Also $\mathrm{IC}(x)$ leads, under certain regularity conditions, to an expression for the asymptotic variance of the statistic $\hat{\theta}_n$, for $n \to \infty$, and asymptotic relative efficiency problems of such statistics may be also studied. The regularity conditions are too complicated to be stated here and are beyond the scope of this work and involve intricacies of functional calculus [cf., Huber (1981), Serfling (1980)]. If these are satisfied, $\sqrt{n}(\hat{\theta}_n - \theta)$ has, for $n \to \infty$, a limiting $N(O, A)$ distribution, where $A = E_F[(\mathrm{IC}(X))^2]$, provided $0 < A < \infty$. [These conditions, however, may be readily stated for some cases as given below.]

For example, for the sample mean $\bar{X}$ and with $\mu = E[X_i]$, we have, by using the fact that $\int x' \, d\delta_x(x') = x$, $\mathrm{IC}(x) = x - \mu$. Hence, $A = E_F[(\mathrm{IC}(X))^2] = \sigma^2$, and the regularity condition, here, is simply that $0 < \sigma^2 < \infty$. We note the sensitivity of the sample mean to the influence of a large observation as $\mathrm{IC}(x) \to \infty$ for $x \to \infty$.

For an M-estimator (§2.25),

$$\mathrm{IC}(x) = \frac{-\psi(x, t_0)}{E_F[(\partial/\partial t_0)\psi(X, t_0)]}$$

and we recover the asymptotic theorem in §2.25(a), and rigorous sufficiency conditions under which the asymptotic theorem holds were stated there. [We note that given an influence curve we may find a ψ-function which is proportional to it.]

The quantity $\sup_x |\mathrm{IC}(x)|$ is called the gross-error sensitivity. [For the mean $\bar{X}$, the latter is infinite.]

Some recommendations on how to choose the ψ-function for an M-estimator (of location) follows:

(1) First in order to have a bounded influence curve and hence a finite gross-error sensitivity, choose $\psi(x; t)$ to be a bounded function of x.
(2) For a symmetric distribution, the objective function (§§1.12, 2.25) should put equal weights to the observations at equal distance from the symmetry point, this leads to the condition $\psi(-x + t) = -\psi(x - t)$.
(3) *Near* the symmetry point, of a symmetric distribution, if the distribution in question is of the form of the normal distribution, then choose $\rho(x - t) \propto (x - t)^2$ for $x \simeq t$. [This last property is called Winsor's principle.]
(4) In order to allow for scale, the objective function is chosen in the form:

$\sum_{i=1}^{n} \rho(u_i)$, where $u_i = (x_i - t)/\hat{S}_n$, and $\hat{S}_n$ is an estimate of scale. $\hat{S}_n$ is chosen in such a way that it is less "sensitive" to outliers than the location estimate itself. This has suggested the use of *median absolute deviation* MAD (as opposed to the standard deviation, for example) as an estimate of scale:

$$\text{MAD} = \text{median}\{|x_i - M|\}, \qquad M = \text{median}\{x_i\}.$$

In order that $\hat{S}_n$ be *consistent* at the normal distribution one then divides MAD by the number 0.6745. That is, one may choose $\hat{S}_n = \text{MAD}/0.6745$. [Note that, as a scale estimate, $\hat{S}_n(x_1 + \lambda, \ldots, x_n + \lambda) = \hat{S}_n(x_1, \ldots, x_n)$ and $\hat{S}_n(\lambda x_1, \ldots, \lambda x_n) = |\lambda| \hat{S}_n(x_1, \ldots, x_n)$ for real λ.]

Based on the above points one may choose (Huber):

$$\psi(u) = \begin{cases} u & \text{if } |u| \leqslant k, \\ k & \text{if } u > k, \\ -k & \text{if } u < -k, \end{cases}$$

where $0 \leqslant k$, $u = (x - t)/\hat{S}_n$, $\hat{S}_n = \text{MAD}/0.6745$. [$k = 0$ corresponds to the median, $k \to \infty$ corresponds to the mean.] One often uses $k \simeq 1.5$.

(iii) *Some Robust Tests for Scales.* Robust tests for the one-sample, two-sample, and k-sample tests on scales (variances) may be carried out by using jackknife statistics (§2.33): For the one-sample problem: H_0: $\sigma^2 = \sigma_0^2$ (fixed), use

$$J = \frac{\sqrt{n}(\tilde{\theta}_. - \theta_0)}{\sqrt{\sum_{i=1}^{n} (\tilde{\theta}_i - \tilde{\theta}_.)^2/(n-1)}};$$

where $\theta_0 = \ln \sigma_0^2$, and $\tilde{\theta}_i$, $\tilde{\theta}_.$ are defined in §2.33(iii). For n sufficiently large (under the condition $0 < \gamma_2 + 2 < \infty$) J has a standard normal distribution, and hence critical values (Z_α, $Z_{\alpha/2}$, etc.) may be obtained directly from tables of the standard normal distribution. The Pitman asymptotic efficiency of the jackknife statistic to the classic (one-sample) chi-square test of scale is *one*. [The latter is also the relative asymptotic efficiency of the jackknife estimator $\tilde{\theta}_.$ to $\ln S^2$ as estimators of $\ln \sigma^2$.]

For the two-sample test of scales: H_0: $\sigma_1^2 = \sigma_2^2$, use

$$J = \frac{(\tilde{\theta}_.^1 - \tilde{\theta}_.^2)}{\sqrt{V_1/n_1 + V_2/n_2}},$$

where $\tilde{\theta}_.^i$ are defined in §2.33 for each ($i = 1, 2$) of the two-samples of sizes n_1 and n_2, respectively, and

$$V_i = \sum_{j=1}^{n_i} (\tilde{\theta}_j^i - \tilde{\theta}_.^i)^2/(n_i - 1), \qquad i = 1, 2.$$

For n_1 and n_2 sufficiently large, and $\gamma_2^1 = \gamma_2^2 \equiv \gamma_2$, $0 < \gamma_2 + 2 < \infty$, J has an asymptotic normal distribution. The Pitman asymptotic efficiency of the jackknife statistic to the classic (two-sample) F-test of scales, based on S_1^2/S_2^2 is *one*. The jackknife method is more powerful than the test based on S_1^2/S_2^2 for the uniform distribution, and for the double-exponential distribution the situation is reversed [Miller (1968)].

For the k-sample problem: H_0: $\sigma_1^2 = \cdots = \sigma_k^2$, choose (Layard (1973))

$$J = \frac{\sum_{i=1}^{k} n_i(\tilde{\theta}_\cdot^i - \tilde{\theta}_\cdot^\cdot)^2/(k-1)}{\sum_{i=1}^{k}\sum_{j=1}^{n_i} (\tilde{\theta}_j^i - \tilde{\theta}_\cdot^i)^2/(n-k)},$$

where $\tilde{\theta}_\cdot^\cdot = \sum_{i=1}^k n_i \tilde{\theta}_\cdot^i/n$, $n = \sum_{i=1}^k n_i$. For $n_1, \ldots, n_k$ not too small, and $\gamma_2^1 = \cdots = \gamma_2^k \equiv \gamma_2$, $0 < \gamma_2 + 2 < \infty$, J has an approximate F-distribution, and, with a level of significance α, one then rejects H_0 if $J > F_\alpha(k-1, n-k)$, where $F_\alpha(v_1, v_2)$ is the $(1-\alpha)$th quantile $P[F \geqslant F_\alpha(v_1, v_2)] = \alpha$ with v_1 and v_2 degrees of freedom, concluding that at least two of the variances $\sigma_1^2, \ldots, \sigma_k^2$ are not equal. The Pitman asymptotic efficiency of the jackknife statistic J to the Bartlett statistic M (§§1.35 and 3.48), for $\sigma_i = \sigma(1 + \theta_i/\sqrt{n})$, $n \to \infty$, is *one*. In general, the test based on the jackknife statistic J is more powerful than the one based on the statistic M for nonnormal populations (Layard).

See also §§1.12, 1.13, 1.35, 1.41, 2.19, 2.25 and 2.33.

[Cf., Huber (1977, 1981), Hoaglin *et al.* (1983), Jaeckel (1971), Hampel (1968, 1971), Bickel (1965), Tukey (1960a, b), Rey (1978, 1983), Lehmann (1963), Miller (1968, 1974), Manoukian (1984a, 1986), Fernholz (1983), Serfling (1980.)]

§1.43. Pitman–Fisher Randomization Methods

Randomization methods are quite appealing as they involve no assumptions on the underlying populations to carry out a test, and the rules for it are formulated, self-consistently, in terms of the original *data* one obtains in the experiment in question. Some properties shared by randomization methods are the following:

(1) The critical values of a test depend on the original data, and hence *no* statistical tables may be prepared, once and for all, to carry out tests, in general, as the critical values will vary from experiment to experiment. Therefore the methods are not usually convenient.
(2) The testing problem reduces to an equivalent one with sampling, *without* replacement, from a *finite* population and involves straightforward though often tedious computations.
(3) As mentioned above, no assumption is made about the underlying populations from which the original data is selected.

(4) By formal substitution of the random variables in the relevant statistics by the *ranks* (see §1.2), or functions thereof, of the observations, one recovers many of the standard nonparametric statistics (§1.44). By doing so, critical values of the tests, in question, may be then tabulated once and for all.

Because of property (1) above, randomization tests are termed as *conditionally nonparametric tests.* When the sample sizes are not small, the computations involved in randomization tests are quite tedious. Although these methods may be of limited applicability, when sample sizes are large, they play an important role in the foundations of nonparametric statistics (§1.44), as mentioned in property (4) above, as they provide the initial steps for the construction of nonparametric methods. For large samples, the distribution of a statistic in a randomization test may be approximated by a well-known distribution (such as the normal or beta distributions) for which statistical table are available and one may then "loose" reference to the original sample. Such approximations are, however, often difficult to assess. From the point of view of power, their asymptotic relative efficiencies are often comparable, and even higher, relative to most powerful parametric tests.

(i) *One-Sample Problem.* Suppose the data consist of a sample $(X_1, Y_1)', \ldots, (X_m, Y_m)'$ of independent random 2-vectors from a bivariate distribution. We consider the *nonzero* differences $(X - Y)$: $Z_1, \ldots, Z_n$ $(n \leqslant m)$. Suppose Z_i has a distribution $P[Z_i \leqslant z] = G(z) = F(z - \theta)$, where θ is a location parameter and will be taken to be the median. We test the hypothesis $H_0\colon \theta = 0$ against $H_A\colon \theta > 0$. By definition of the median, the distribution of each Z_i is symmetric about 0 under the null hypothesis. Let $z_1, \ldots, z_n$ denote the *original data* restricted to the nonzero values. We note that $z_1, \ldots, z_n$ are fixed (nonzero) numbers. Define (see §3.10) the collection of all 2^n n-vectors:

$$\Omega = \{(+|z_1|, \ldots, +|z_n|)', (-|z_1|, +|z_2|, \ldots, +|z_n|)', \ldots, (-|z_1|, -|z_2|, \ldots, -|z_n|)'\}$$

by attaching + or − signs, in all possible ways, to $|z_1|, \ldots, |z_n|$. Let $\mathbf{S} = (S_1, \ldots, S_n)'$ be a random vector taking values from Ω with equal probabilities. *If* the Z_i were distributed normally with unknown variance, then the uniformly most powerful unbiased test is obtained by rejecting H_0 for large values of the Student statistic:

$$T = \frac{\sqrt{n}\,\bar{Z}}{\sqrt{\sum_{i=1}^{n} (Z_i - \bar{Z})^2/(n-1)}}.$$

If one formally replaces each Z_i by S_i in T, then one notes that the denominator in T is positive and decreases as $\bar{S}$ increases. The T-statistic then suggests choosing the statistic $S = \sum_{i=1}^{n} S_i$ to carry out the test. Under the null hypothesis, all the "outcomes" in Ω should have occurred with equal probabilities thus suggesting to reject the null hypothesis for large values of S. The distribution of S is discussed in §3.10, and the first four moments of S are also given there. Let $\sum_{i=1}^{k} z_i = c$ and let q be the number of such $S \geqslant c$. That is, $P[S \geqslant c] = q/2^n$. If the level of significance of the α-test is such that $q/2^n \leqslant \alpha$, then reject the null hypothesis H_0: $\theta = 0$. If the second term in the square brackets in $E[S^4]$, given in §3.10, is small in comparison to 3, then $S/\sqrt{E[S^2]}$ has approximately the first four moments (also the odd ones) of a standard normal distribution.

(ii) *Pitman's Permutation Test of The Correlation Coefficient.* Suppose $(X_1, Y_1), \ldots, (X_n, Y_n)$ is a sample of size n from a bivariate distribution with correlation coefficient ρ. If the underlying distribution is the bivariate normal one, then the uniformly most powerful unbiased test rejects the hypothesis H_0: $\rho = 0$, against H_A: $\rho > 0$ for large values of the sample correlation coefficient

$$R = \frac{\sum_{i=1}^{n} (X_i - \bar{X})(Y_i - \bar{Y})}{\sqrt{\sum_{i=1}^{n} (X_i - \bar{X})^2 \sum_{i=1}^{n} (Y_i - \bar{Y})^2}}.$$

We now consider the test based on the randomization method. We wish to test the hypothesis: H_0: X and Y are independent, against, for example, H_A: large values of X have the tendency to be paired with large values of Y. Let $(x_1, y_1), \ldots, (x_n, y_n)$ denote the original data and, without loss of generality, arranged in such a manner that $\bar{x} = 0$, $\bar{y} = 0$ by shifting the values of x_i and y_i. We fix the order $x_1, \ldots, x_n$ and we consider (§3.11) all $n!$ pairings of the y's with the x's to generate the collection of all $n!$ n-tuples:

$$\Omega = \{((x_1, y_1), \ldots, (x_n, y_n))', ((x_1, y_2), (x_2, y_1), \ldots, (x_n, y_n))', \ldots,$$
$$((x_1, y_n), \ldots, (x_n, y_1))'\}.$$

Let $((x_1, W_1), \ldots, (x_n, W_n))'$ be a random variable taking any of the $n!$ values from Ω with equal probabilities. If we formally replace the X_i and Y_i in R by x_i and W_i, respectively, we note that the denominator of R is a constant number. This suggests defining the statistic $S = \sum_{i=1}^{n} x_i W_i$. Under the null hypothesis all the $n!$ pairings in Ω should have occurred with equal probabilities thus suggesting to reject the null hypothesis for large values of S. The distribution of S, as well as its four moments, are discussed in §3.11. Let $\sum_{i=1}^{n} x_i y_i = c$ and let q be the number of S-values $\geqslant c$. That is, $P[S \geqslant c] = q/n!$. If the level of significance α of the test is such that $\alpha \geqslant q/n!$ then reject the null hypothesis.

(iii) *Two-Sample Problem.* Let $X_1, \ldots, X_{n_1}$ and $Y_1, \ldots, Y_{n_2}$ be two samples from two populations, such that each X and Y have distributions $F(x)$ and

$F(x - \theta)$, respectively. Here θ is a location parameter. We test the hypothesis H_0: $\theta = 0$ against H_A: $\theta > 0$. *If* the underlying populations were normal with means μ_1 and μ_2, respectively, and a common variance, then the uniformly most powerful unbiased test, with $\mu_2 - \mu_1 \equiv \theta$, is obtained by rejecting H_0 for large values of the Student statistic:

$$T = \frac{\bar{Y} - \bar{X}}{\sqrt{\dfrac{1}{n_1} + \dfrac{1}{n_2}}} \Bigg/ \sqrt{\frac{\sum_{i=1}^{n_1} (X_i - \bar{X})^2 + \sum_{i=1}^{n_2} (Y_i - \bar{Y})^2}{(n_1 + n_2 - 2)}}$$

of $(n_1 + n_2 - 2)$ degrees of freedom. Let $x_1, \ldots, x_{n_1}, y_1, \ldots, y_{n_2}$ denote the original data. We rewrite the latter as $x_1, \ldots, x_{n_1}, x_{n_1+1}, \ldots, x_{n_1+n_2}, n_1 + n_2 = N$. Generate (§3.12) the collection $\Omega = \{(z_1, \ldots, z_{n_2})\}$ of all n_2-tuples $(z_1, \ldots, z_{n_2})$, $n_2 < N$, which may be selected from $(x_1, \ldots, x_N)$ such that all orderings of the elements in $(z_1, \ldots, z_{n_2})$ are considered as equivalent. Clearly, Ω contains $\binom{N}{n_2}$ n_2-tuples. Let $(Z_1, \ldots, Z_{n_2})$ denote a random variable taking values from Ω with equal probabilities. Given $Z_1, \ldots, Z_{n_2}$, the "remaining" $Z_{n_2+1}, \ldots, Z_n$ from $(x_1, \ldots, x_N)$ are uniquely determined (up to an ordering). If we formally replace $Y_1, \ldots, Y_{n_2}$ and $X_1, \ldots, X_{n_1}$ in T by $Z_1, \ldots, Z_{n_2}$ and $Z_{n_2+1}, \ldots, Z_N$, then using the fact that $(\sum Z_i + \sum W_i)$ and $(\sum Z_i^2 + \sum W_i^2)$ are con stants we are led to consider the statistic $S = \sum_{i=1}^{n_2} Z_i$. Under the null hypothesis all the n_2-tuples in Ω "should have occurred" with equal probabilities, suggesting to reject H_0 whenever S takes on large values. The distribution, and some of the moments of S, are discussed in §3.12. Let $\sum_{i=n_1+1}^{N} x_i = c$, and let q be the number of S-values $\geqslant c$. That is, $P[S \geqslant c] = q \Big/ \binom{N}{n_2}$. Then with a level of significance α, reject H_0 if $q \Big/ \binom{N}{n_2} \leqslant \alpha$.

(iv) *k-Sample Problem.* Suppose we have at our disposal k populations under k different treatments. Samples of sizes $n_1, \ldots, n_k$ are, respectively, chosen from these populations. Let X_{ij} denote the random variable representing the jth observation in the ith sample. We denote the original data by $x_{11}, \ldots, x_{1n_1}, x_{21}, \ldots, x_{2n_2}, \ldots, x_{k1}, \ldots, x_{kn_k}, \sum_{i=1}^{k} n_i = n$. Let (§3.13) Ω denote the collection of all partitions of the set of numbers $x_{11}, \ldots, x_{kn_k}$ into k groups of sizes $n_1, \ldots, n_k$, respectively. The elements in Ω will be denoted by $(\mathbf{z}_1, \ldots, \mathbf{z}_k)$, where $\mathbf{z}_i$ is n_i-dimensional, and all permutations of the components of a $\mathbf{z}_i$ are considered as equivalent. Clearly, Ω contains $n!/n_1! \ldots n_k!$ elements. Let $(\mathbf{Z}_1, \ldots, \mathbf{Z}_k)$ denote a random variable taking values from Ω with equal probabilities:

$$P[(\mathbf{Z}_1, \ldots, \mathbf{Z}_k) = (\mathbf{z}_1, \ldots, \mathbf{z}_k)] = n_1! \ldots n_k!/n!, \qquad (\mathbf{z}_1, \ldots, \mathbf{z}_k) \in \Omega.$$

The classic one-way analysis of variance statistic (§1.40) is given by

$$F = \frac{\left[\sum_{i=1}^{k} n_i(\bar{X}_{i.} - \bar{X}_{..})^2/(k-1)\right]}{\left[\sum_{i=1}^{k}\sum_{j=1}^{n_i} (X_{ij} - \bar{X}_{i.})^2/(n-k)\right]},$$

where

$$\bar{X}_{..} = \sum_{i=1}^{k}\sum_{j=1}^{n_i} X_{ij}/n, \qquad \bar{X}_{i.} = \sum_{j=1}^{n_i} X_{ij}/n_i,$$

and one rejects the hypothesis of the *equivalence of the treatments* for large values of F. We write $\mathbf{Z}_i = (Z_{i1}, \ldots, Z_{in_i})'$. If we formally replace X_{ij} by Z_{ij} in F, then by noting that $\sum_{i=1}^{k}\sum_{j=1}^{n_i} Z_{ij}$ and $\sum_{i=1}^{k}\sum_{j=1}^{n_i} Z_{ij}^2$ are constants we are led to consider the statistic $S = \sum_{i=1}^{k} n_i \bar{Z}_{i.}^2$, where $\bar{Z}_{i.} = \sum_{j=1}^{n_i} Z_{ij}/n_i$. The distribution of S and some of its moments are discussed in §3.13. Let $\sum_{i=1}^{k} n_i \bar{x}_{i.}^2 = c$, and let q be the number of S-values $\geqslant c$. That is, $P[S \geqslant c] = qn_1!\ldots n_k!/n!$. Then with a level of significance α, reject the hypothesis of the equivalence of the k treatments if $qn_1!\ldots n_k!/n! \leqslant \alpha$.

See also §§1.40, 1.44 and 3.10–3.13.

[Cf., Pitman (1937a, b, c), Fisher (1951), Bradley (1968), Manoukian (1985), Edgington (1980).]

§1.44. Nonparametric Methods

A *nonparametric* statistical procedure may be defined as any statistical procedure which has certain properties holding true under very few assumptions made about the underlying populations from which the data are obtained. By a *distribution-free inference*, it is meant that the distribution of the statistic, on which the inference is based, is independent of the specific distribution of the underlying populations and is the same for all distributions in some class of distributions. By a distribution-free test it is usually meant that the distribution of the statistic in question, under the null hypothesis (and hence also the significance level of the test), is the same for all distributions in some well-defined class C of distributions containing more than one distribution.

Nonparametric statistical methods are attractive not only because they involve very few assumptions in their formulations, but because they are also often very easy to apply and may not require detailed complicated data. As a matter of fact, many statistics are based on the ranks (§1.2) of the observations and hence do not require explicit numerical values associated with the observations. If the actual values of the observations are available and one uses a rank test statistic to formulate a test of hypothesis, by discarding in the process the actual numerical values of the observations, one may expect a drastic loss of efficiency relative to a classical counterpart test, at a given model, which makes use of this additional information. Surprisingly, this is not generally the case and quite often the efficiency loss of a nonparametric test relative to a classical counterpart is only slight for normal populations, and the former is usually more efficient than the latter for nonnormal populations.

(A) One-Sample Problems

(i) *The Sign Test.* Perhaps one of the simplest nonparametric test when it comes to ease of computation is the sign test. Let $Z_1, \ldots, Z_n$ be independent identically distributed random variables each with a continuous distribution $G(z) \equiv F(z - \theta)$, where θ is a location parameter and will be taken to be the median. By definition of the median, $P[Z_i > \theta] = P[Z_i < \theta] = \frac{1}{2}$. [Also note that due to the continuity assumption of the distribution $P[Z_i = 0] = 0$.] Consider the test $H_0\colon \theta = 0$ against $H_A\colon \theta > 0$. Clearly a large fraction of positive Z_i is an indication that H_0 is false. Accordingly, we introduce the statistic $B = \sum_{i=1}^{n} \psi(Z_i)$, where

$$\psi(x) = 1 \quad \text{if } x > 0, \qquad \psi(x) = 0 \quad \text{if } x < 0.$$

Let $p = P[Z_i > 0]$. Under the null hypothesis, $p = \frac{1}{2}$. With a level of significance α, H_0 is rejected if $B \geqslant b_c$, where b_c is the smallest positive integer such that

$$(\tfrac{1}{2})^n \sum_{k=b_c}^{n} \binom{n}{k} \geqslant \alpha.$$

The power of the test is given by

$$\mathscr{P}(\theta) = \sum_{k=b_c}^{n} \binom{n}{k} (p)^k (1 - p)^{n-k}.$$

The test is unbiased, and for $p > \frac{1}{2}$ it is consistent. To ensure that $p > \frac{1}{2}$ it is sufficient to assume that $F(x)$ is also strictly increasing. If $F(x)$ is also symmetric, that is $F(-x) = 1 - F(x)$, then the median coincides with the mean. The Student t-test then rejects H_0 for large values of the statistic $T = \sqrt{n}\,\bar{Z}/S$. If, for example, $0 < \gamma_2 + 2 < \infty$, then the Pitman asymptotic efficiency of the sign test to the t-test is given by $e_\infty(B, T) = 4\sigma^2 f(0)$, where σ^2 is the variance of Z_i, and $f(x)$ is the density associated with the continuous (here also symmetric) distribution $F(x)$. For a normal population, $e_\infty(B, T) = 2/\pi$. [For handling ties in the observations see, for example, Putter (1955), Bradley (1969), Emerson and Simon (1979).]

(ii) *Wilcoxon Signed Rank Test.* In the sign test, we have merely needed the signs of the Z_i; that is, whether $Z_i > 0$ or $Z_i < 0$. In the Wilcoxon signed rank test, we also make use of the information (if available) of the relative magnitudes $|Z_i|$ of the Z_i. We also assume that the continuous distribution $F(x)$ is symmetric; that is, $F(x) = 1 - F(-x)$. A statistic equivalent to S in the one-sample randomization test (i), §1.43, is

$$S' = \sum_{i=1}^{n} \frac{(\text{sign } S_i + 1)}{2} |S_i|,$$

since $\sum_{i=1}^{n} |S_i|$ is a constant, and

$$\text{sign } S_i = +1 \quad \text{if } S_i > 0, \quad \text{and} \quad \text{sign } S_i = -1 \quad \text{if } S_i < 0.$$

If we *formally* replace $|S_i|$ by the rank R_i of $|Z_i|$ in $(|Z_1|, \ldots, |Z_n|)$ and define sign $Z_i = 0$ if $Z_i = 0$, then we obtain the Wilcoxon one-sample statistic:

$$W = \sum_{i=1}^{n} \frac{(\text{sign } Z_i + 1)}{2} R_i.$$

The null distribution of W is given in §3.7. and is clearly distribution-free within the class of distributions defined above. For $\theta > 0$, we expect that the Z_i take on "large" positive values and few take on negative values with large absolute values. Accordingly, we reject the hypothesis $H_0\colon \theta = 0$ against $H_A\colon \theta > 0$, for large values of W.

The mean and the variance of the statistic W are given by:

$$E[W] = \frac{n(n-1)}{2} p_1 + p,$$

$$\sigma^2(W) = \frac{n(n-1)}{2} p_1(1 - p_1) + n(n-1)(n-2)(p_2 - p_1^2)$$
$$+ 2n(n-1)(p_3 - p_1 p_2) + np(1-p),$$

where

$$P[Z_1 > 0] \equiv p, \qquad P[Z_1 + Z_2 > 0] \equiv p_1,$$
$$P[Z_1 + Z_2 > 0, Z_1 + Z_3 > 0] \equiv p_2, \qquad P[Z_1 + Z_2 > 0, Z_1 > 0] \equiv p_3.$$

Theorem. *If $p_2 - p_1^2 > 0$, then $(T_n - E[T_n])/\sigma(T_n)$ has, for $n \to \infty$, a limiting $N(0, 1)$ distribution, where $T_n = W/n(n-1)$. The test is consistent if $p_1 > \frac{1}{2}$. [To achieve this it is sufficient to assume that $F(x)$ is also strictly increasing.] If $0 < \gamma_2 + 2 < \infty$, where γ_2 is the kurtosis of the distribution F, then the Pitman asymptotic efficiency of W to the classic t-test is*

$$e_\infty(W, T) = 12\sigma^2 \left(\int_{-\infty}^{\infty} f^2(x)\, dx \right)^2,$$

where $f(x)$ is the density associated with $F(x)$, and σ^2 is the variance of the Z_i. We note that (Hodges–Lehmann bound) $e_\infty(W, T) \geqslant \frac{108}{125} \simeq 0.864$. For a normal distribution, $e_\infty(W, T) = 3/\pi \simeq 0.955$.

The development of an *optimum* test for all $\theta \geqslant 0$ based on (sign $Z_1, \ldots,$ sign Z_n; $R_1, \ldots, R_n$) whose corresponding distribution may be evaluated is rarely possible in practice, and one develops instead a locally (§1.33) most powerful test for testing H_0: $\theta = 0$ against H_A: $\theta > 0$ (for small θ).

Theorem. *A locally most powerful test is obtained by rejecting H_0 for large values of the statistic:*

$$-\sum_i (\text{sign } Z_i) E\left[\frac{f'(H_{(R_i)})}{f(H_{(R_i)})}\right], \tag{$*$}$$

where $f'(x) = (d/dx)f(x)$, and $H_{(1)} < \cdots < H_{(n)}$ is an ordered sample from a distribution with density:

$$g(h) = \begin{cases} 2f(h), & h > 0, \\ 0, & h < 0, \end{cases}$$

and we recall $f(x)$ is the density associated with (continuous and symmetric) $F(x)$, where $Z_1, \ldots, Z_n$ is a sample from the distribution $G(z) \equiv F(z - \theta)$.

If

$$f(x) = \frac{1}{\sqrt{2\pi}\,\sigma} \exp[-x^2/2\sigma^2],$$

then the expectation value in ($*$) reduces to $E[H_{(R_i)}]$.

For a *logistic distribution* (§3.27) with

$$f(x) = e^x[1 + e^x]^{-2},$$

the statistic in ($*$) becomes

$$\sum_i (\text{sign } Z_i) R_i/(n + 1)$$

or equivalently it reduces to the Wilcoxon one-sample statistic W, since $\sum_i R_i$ is a constant. Thus we *recover* the Wilcoxon signed rank statistic, as a special case, and the latter provides a locally most powerful test for detecting a shift in location for the logistic distribution.

A *power comparison* [Randles and Wolfe (1979)] based on Monte Carlo studies shows, in general, that the t-test is superior to the W-test for the uniform and normal distributions and the two tests are superior to the sign test for these distributions. For the logistic distribution, the t- and W-tests are comparable and the two are more powerful than the sign test. For the Cauchy distribution, the sign test is more powerful than the W-test and the two tests are more powerful than the t-test.

(iii) *Kendall's Test.* Let $(X_1, Y_1), \ldots, (X_n, Y_n)$ denote a sample from a bivariate distribution which together with the marginal distributions are continuous. We say that the pairs (X_i, X_j) and (Y_i, Y_j), $i \neq j$, are concordant if $(X_i - X_j)(Y_i - Y_j) > 0$ and discordant if $(X_i - X_j)(Y_i - Y_j) < 0$. In the first case, large (small values of X have the tendency to be paired with large (small) values of Y. In the second case, large (small) values of X have the tendency to be paired with small (large) values of Y. As a measure of "association" between X and Y, we define the parameter τ:

$$\tau = P[(X_2 - X_1)(Y_2 - Y_1) > 0] - P[(X_2 - X_1)(Y_2 - Y_1) < 0].$$

Clearly, if X and Y are independent then $\tau = 0$. In general, $-1 \leqslant \tau \leqslant 1$. The basic idea in developing the Kendall test is to find an unbiased estimator of τ. Such a statistic is defined by:

$$K = \frac{2}{n(n-1)} \sum_{i<j} \psi(X_i, X_j; Y_i, Y_j),$$

where

$$\psi(x_1, x_2; y_1, y_2) = \begin{cases} 1 & \text{for } (x_2 - x_1)(y_2 - y_1) > 0, \\ 0 & \text{for } \qquad\qquad\qquad\qquad = 0, \\ -1 & \text{for } \qquad\qquad\qquad\qquad < 0. \end{cases}$$

The statistic K is referred to the Kendall rank correlation coefficient, and $E[K] = \tau$. For $\tau = 0$, $\sigma^2(K) = 2(2n + 5)/9n(n - 1)$, and $K/\sigma(K)$ has, for $n \to \infty$, a limiting $N(0, 1)$ *distribution.*

We are interested in the tests H_0: X and Y are independent and hence $\tau = 0$ against

H_A: pairs of observations have the tendency to be concordant,

i.e., $\tau > 0$; or against

H'_A: pairs of observations have the tendency to be discordant,

i.e., $\tau < 0$; or against

H''_A: pairs of observations have the tendency to be either concordant or have the tendency to be discordant,

i.e., $\tau \neq 0$. The null hypothesis H_0 is then rejected for large, or for small, or for neither small nor large values of K, respectively. These tests are consistent for $\tau > 0$, $\tau < 0$, $\tau \neq 0$, respectively.

For a bivariate normal distribution, the Pitman asymptotic efficiency of the K-statistic to the correlation coefficient R-statistic (§3.50), with $\rho = O(n^{-1/2})$, which in turn implies that $\tau = O(n^{-1/2})$, is given by $9/\pi^2$.

(iv) *Spearman's Test.* Let $(X_1, Y_1), \ldots, (X_n, Y_n)$ denote a sample from a bivariate distribution which together with the marginal distributions are continuous. As in Kendall's test, we are interested in testing: H_0: the X and Y are independent, against H_A: large values of X have a tendency to be paired with larger values of Y, or against H'_A: larger values of X have a tendency to be paired with smaller Y-values and vice versa, or against H''_A: either there is a tendency that larger values of X are paired with larger values of Y or there is a tendency that larger values of X are paired with smaller Y-values. Let R_i denote the rank of X_i in $\{X_i, \ldots, X_n\}$, and T_i denote the rank of Y_i in $\{Y_1, \ldots, Y_n\}$. The Spearman statistic is defined by

$$S = \frac{\sum_{i=1}^{n} (T_i - \bar{T}_.)(R_i - \bar{R}_.)}{\sqrt{\sum_{i=1}^{n} (T_i - \bar{T}_.)^2 \sum_{j=1}^{n} (R_i - \bar{R}_.)^2}},$$

where

$$\bar{R}_. = \sum_{i=1}^{n} R_i/n = (n+1)/2, \qquad \bar{T}_. = \sum_{i=1}^{n} T_i/n = (n+1)/2.$$

The denominator in S is a constant, and the statistic S may be written in the equivalent form:

$$S = 1 - \frac{2(2n+1)}{(n-1)} + \frac{2}{n(n^2-1)} \sum_{i=1}^{n} R_i T_i,$$

which is equivalent to the statistic $\sum_{i=1}^{n} R_i T_i$. If we formally replace x_i and W_i, respectively, by R_i and T_i in S in Pitman's permutation test of the correlation coefficient (§1.43(ii)) we *obtain* this statistic. For the above-mentioned tests, the null hypothesis is rejected for large values of S, for small values of S and for S not too large or not too small, respectively. To find the null distribution, we note that without loss of generality, we may suppose that the Y-values are arranged in ascending order $Y_1 < \cdots < Y_n$. The X-values are then labelled accordingly, and we may write $\sum_{i=1}^{n} R_i T_i$ simply as $\sum_{i=1}^{n} iR_i$. Under the null hypothesis, all possible $n!$ rankings $R_1, \ldots, R_n$ occur with equal probabilities. If we denote by $c(n, s)$ the number of S obtained from such $n!$ rankings such that $S = s$, then $P_{H_0}[S = s] = c(n, s)/n!$. Also $E_{H_0}[S] = 0$, $\sigma^2_{H_0}(S) = (n-1)^{-1}$, and S is symmetrically distributed about the origin. For methods of dealing with ties in the observations; see, e.g., Kendall (1975).

(v) *Kolmogorov–Smirnov Tests.* Let X be a random variable with unknown continuous distribution function $G(x)$. We are interested in the following tests: H_0: $G(x) \equiv F(x)$, against H_A: $G(x) \not\equiv F(x)$ or against H'_A: $G(x) > F(x)$, for at least one x, or against H''_A: $G(x) < F(x)$, for at least one x, where $F(x)$ is a completely specified continuous distribution. The Kolmogorov–Smirnov statistics are defined, respectively, by

$$D_n = \sup_x |\hat{F}_n(x) - F(x)|,$$

$$D_n^+ = \sup_x (\hat{F}_n(x) - F(x)),$$

$$D_n^- = \sup_x (F(x) - \hat{F}_n(x)),$$

where $\hat{F}_n(x)$ is the empirical distribution:

$$\hat{F}_n(x) = [\text{number of } X_i \leqslant x]/n,$$

$X_1, \ldots, X_n$ are independent identically distributed random variables each with distribution $G(x)$. The null hypothesis H_0 is then rejected for large values of these statistics, respectively. The null distributions of these statistics are given in §3.14. We recall, according to the Glivenko–Cantelli theorem §2.21, if $G(x) \equiv F(x)$, $\sup_x |\hat{F}_n(x) - F(x)|$ converges with probability one to zero, and, in particular (§2.20), $\hat{F}_n(x)$ converges in probability to $F(x)$, for $n \to \infty$. Under the null hypothesis H_0, and if $F(x)$ is a continuous distribution, the statistics D_n, D_n^+, D_n^- are distribution-free. The asymptotic null distributions of these statistics are given in §2.22(i). Let

$$\Delta_1 = \sup_x |F(x) - G(x)|,$$

$$\Delta_2 = \sup_x (G(x) - F(x)),$$

and

$$\Delta_3 = \sup_x (F(x) - G(x)).$$

Then these tests are consistent if $\Delta_1 > 0$, $\Delta_2 > 0$ and $\Delta_3 > 0$, respectively, for all continuous distributions $F(x)$ and $G(x)$. The statistic D_n is referred to as the Kolmogorov statistic. We note from the very definition of D_n, this statistic often lacks a sensitivity to departures in the *tails* of the distribution (this is also true for D_n^+, D_n^- as well). In this respect a class of statistics called Cramér–von Mises statistics [see e.g., Durbin (1973), Mason and Schuenemeyer (1983)] put more weights on the tails and hence remedy this problem.

A comparison of the D_n-test and the Pearson χ^2-test (§2.23) of fit

(1) The D_n-test may be applied only for continuous distributions, while the χ^2-test applies to both discrete and continuous distributions. As a matter of fact, for the case of discrete distributions, the actual size of the test α_D is such that $\alpha_D \leqslant \alpha$, where α is the nominal size of the test corresponding to tabulated critical values based on a continuity assumption, thus demonstrating a "conservative" character of the D_n-test.
(2) The exact critical values of the D_n-test are known, while for the χ^2-test, one uses the approximate critical values as obtained from the chi-square distribution (§2.23), obtained in the limit $n \to \infty$. In general, however [Slaketer (1965, 1966)], the χ^2-test is more valid than the D_n-test, where validity is a measure of discrepancy between the nominal size (tabled) and the actual size of the test.
(3) The Kolmogorov D_n-test seems to be more powerful than the χ^2-test for any sample size [Massey (1950, 1951a, b), Lilliefors (1967)].

(vi) *Hodges–Lehmann Estimators of a Location Parameter* θ. These are given in §1.15 based on the sign test and Wilcoxon signed rank test statistics.

(B) Two-Sample Problems

(i) *Wilcoxon–Mann–Whitney Test.* Let $X_1, \ldots, X_{n_1}$ and $Y_1, \ldots, Y_{n_2}$ be independent random variables such that each of the X_i has a continuous distribution $F(x)$ and each of the Y_i has a continuous distribution $G(x)$.

We consider the test: H_0: $G(x) = F(x)$ against H_A: $G(x) = F(x - \theta)$, where $\theta > 0$. Let $R_1, \ldots, R_{n_2}$ denote the ranks of $Y_1, \ldots, Y_2$ in the combined sample $\{X_1, \ldots, X_{n_1}, Y_1, \ldots, Y_{n_2}\}$. If we formally replace the Z_i in the two-sample randomization test (§1.43) in the statistic S by R_i we obtain the Wilcoxon statistic: $W_0 = \sum_{i=1}^{n_2} R_i$. The null distribution of W_0 is given in §3.8. The null hypothesis is rejected for large values of W_0. The Mann–Whitney statistic is defined by

$$U = W_0 - \frac{n_2(n_2 + 1)}{2}.$$

Let

$$p_1 = \int_{-\infty}^{\infty} F(x)\, dG(x),$$

$$p_2 = \int_{-\infty}^{\infty} F^2(x)\, dG(x),$$

$$p_3 = \int_{-\infty}^{\infty} [1 - G(x)]^2\, dF(x).$$

Then

$$E[U] = n_1 n_2 p_1,$$

and

$$\sigma^2(U) = n_1 n_2 [p_1 - p_2 - p_3 + p_1^2 + n_1(p_2 - p_1^2) + n_2(p_3 - p_1^2)].$$

Theorem. *Let* $N \to \infty$, $n_1/N \to \lambda$, $n_2/N \to 1 - \lambda$, $0 < \lambda < 1$, *where* $N = n_1 + n_2$. *Then* $(U - E[U])/\sigma(U)$ *has a limiting* $N(0, 1)$ *distribution provided* $\lambda(p_2 - p_1^2) + (1 - \lambda)(p_3 - p_1^2) > 0$. [*Under the null hypothesis* $\lambda(p_2 - p_1^2) + (1 - \lambda)(p_3 - p_1^2) = \frac{1}{12}$.]

The test is *consistent* if $p_1 > \frac{1}{2}$. To achieve this it is sufficient to assume that $G(x)$ is also strictly increasing.

Now suppose

$$E[Y] - E[X] = \theta, \qquad \sigma^2(X) = \sigma^2(Y) \equiv \sigma^2, \qquad \gamma_2^{(1)} = \gamma_2^{(2)} \equiv \gamma_2,$$

$0 < \gamma_2 + 2 < \infty$. Then the Pitman asymptotic efficiency of the W_0-test to the two-sample Student t-test is given by

$$e_\infty(W_0, T) = 12\sigma^2\left(\int_{-\infty}^{\infty} f^2(x)\,dx\right)^2 \qquad \left(\geqslant \tfrac{108}{125} \simeq 0.864\right),$$

where $f(x)$ is the density associated with the distribution $F(x)$. For the normal distribution, $e_\infty(W_0, T) = 3/\pi$.

A *power comparison* [Randles ad Wolfe (1979)] of the W_0- and the t-tests, based on Monte Carlo studies, yields the following general conclusions: For the uniform, the normal, and logistic distributions the two tests are comparable. For the double exponential, the exponential, and the Cauchy distributions, the W_0-test is superior to the t-test.

Theorem. *A locally most powerful test for testing H_0: $\theta = 0$ against H_A: $\theta > 0$, (that is, for small positive θ) based on the ranks of $X_1, \ldots, X_{n_1}, Y_1, \ldots, Y_{n_2}$ in the combined sample $\{X_1, \ldots, X_{n_1}, Y_1, \ldots, Y_{n_2}\}$ is obtained by rejecting the null hypothesis for large values of the statistic:*

$$-\sum_{i=1}^{n_2} E\left[\frac{f'(H_{(R_i)} - \theta)}{f(H_{(R_i)})}\right], \tag{$*$}$$

where $H_{(1)} < \cdots < H_{(N)}$ is an ordered sample of size $n_1 + n_2 = N$ from a distribution with density $f(x)$. [The expectation $E[\cdot]$ is with respect to the distribution $F(x)$.] If $f(x)$ is the probability density function of the standard normal distribution then the statistic in $(*)$ becomes

$$\sum_{i=1}^{n_2} E[H_{(R_i)}],$$

where $H_{(1)} < \cdots < H_{(N)}$ is an ordered sample from a standard normal distribution. The latter statistic is then called the normal scores statistic and is sometimes referred to as the Fisher–Yates–Hoeffding–Terry statistic. It may be also rewritten as

$$\sum_{i=1}^{n_2} E[\phi^{-1}(U_{(R_i)})],$$

where $\phi(x)$ is the standard normal distribution, and $U_{(1)} < \cdots < U_{(N)}$ is an ordered sample from a uniform distribution over (0, 1). *This statistic has the remarkable property of having a minimum asymptotic relative efficiency of* one *compared to the classic t-test. By construction, it yields the locally most powerful test, based on the ranks of the observations in the pooled sample, for detecting a shift in the mean of the normal distribution. For a logistic distribution, the statistic in* $(*)$ *is equivalent to the Wilcoxon statistic, and we* recover *this statistic as a special case of the one in* $(*)$ *for such an alternative. Thus the Wilcoxon–Mann–Whitney statistic provides the locally most powerful test, based on the ranks of the observations of the pooled sample, for detecting a shift in location of a logistic distribution.*

(ii) *Kolmogorov–Smirnov Tests.* Let $X_1, \ldots, X_{n_1}$ and $Y_1, \ldots, Y_{n_2}$ be independent random variables such that each of the X's has a continuous distribution $F(x)$ and each of the Y's has a continuous distribution $G(x)$. We are interested in the tests: H_0: $F(x) \equiv G(x)$ against H_A: $F(x) \not\equiv G(x)$, or against H'_A: $F(x) > G(x)$. The Kolmogorov–Smirnov statistics for these tests are, respectively, defined by

$$D_{n_1,n_2} = \sup_x |\hat{F}_{n_1}(x) - \hat{G}_{n_2}(x)|,$$

$$D^+_{n_1,n_2} = \sup_x (\hat{F}_{n_1}(x) - \hat{G}_{n_2}(x)),$$

where

$$\hat{F}_{n_1} = \frac{[\text{number of } X_i \leqslant x]}{n_1},$$

$$\hat{G}_{n_2} = \frac{[\text{number of } Y_i \leqslant x]}{n_2}.$$

The null distributions of D_{n_1,n_2} and $D^+_{n_1,n_2}$ (also of $D^-_{n_1,n_2}$ with corresponding alternative H''_A: $F(x) < G(x)$), for the case $n_1 = n_2 = n$, are given in §3.15. The asymptotic null distributions of these statistics are given in §2.22(ii). If we define

$$\Delta_1 = \sup_x |F(x) - G(x)|, \qquad \Delta_2 = \sup_x (F(x) - G(x)),$$

then these tests are consistent, respectively, for $\Delta_1 > 0$, $\Delta_2 > 0$.

(C) k-Sample Problems

(i) *Kruskal–Wallis One-Way Layout Test.* Let $\{X_{ij}\}$, for $j = 1, \ldots, n_i$, $i = 1, \ldots, k$, be a set of independent random variables, and let $F_i(x)$ denote the continuous distribution of the X_{ij}, $j = 1, \ldots, n_i$. We consider the model

$$X_{ij} = \mu + \alpha_i + \varepsilon_{ij},$$

where μ is the overall mean, and the α_i are treatment effects, $\sum_{i=1}^k n_i\alpha_i = 0$. The parameters $(\mu + \alpha_i)$ are location parameters, and the errors ε_{ij} are assumed to be independent identically distributed. We assume that $F_i(x) = G(x - \alpha_i)$.

The hypothesis to be tested is that there is no difference between the treatment effects. Accordingly we test the hypothesis H_0: $\alpha_1 = \cdots = \alpha_k$ against H_A: $\alpha_i \neq \alpha_j$ for at least a pair (i, j), $i \neq j$ of integers in $[1, \ldots, k]$.

Let R_{ij} denote the rank of X_{ij} in the pooled sample $\{X_{11}, \ldots, X_{1n_1}, \ldots, X_{k1}, \ldots, X_{kn_k}\}$. Define

$$\bar{R}_{i.} = \sum_{j=1}^{n_i} R_{ij}/n_i, \qquad N = \sum_{i=1}^{k} n_i.$$

The Kruskal–Wallis statistic is defined by:

$$H = \frac{12}{N(N+1)} \sum_{i=1}^{k} n_i \left(\bar{R}_{i.} - \frac{N+1}{2} \right)^2.$$

If the null hypothesis is false, then H will tend to take large values, and we reject H_0 in favor of H_A. Under the null hypothesis the pooled sample $\{X_{11}, \ldots, X_{1n_1}, \ldots, X_{k1}, \ldots, X_{kn_k}\}$ may be thought to have been selected from the same population. Hence, if $c(n_1, \ldots, n_k, h)$ denotes the number of H obtained from the $N!/n_1! \ldots n_k!$ partitions of the N observations into all possible k subgroups of sizes $n_1, \ldots, n_k$ such that $H = h$, then

$$P[H = h] = \frac{c(n_1, \ldots, n_k, h) n_1! \ldots n_k!}{N!}.$$

The statistic H may be also rewritten as

$$H = [12/N(N+1)] \sum_{i=1}^{k} n_i \bar{R}_{i.}^2 - 3(N+1),$$

which is equivalent to the statistic $\sum_{i=1}^{k} n_i \bar{R}_{i.}^2$. The latter *coincides* with the S-statistic in the k-sample randomization test in §1.43, when the Z_{ij} are replaced by the ranks R_{ij} in S. Again by doing so one "looses" reference, in the randomization method, to the sample and one obtains a nonparametric test.

Theorem. *Let* $\alpha_i = \theta_i/\sqrt{N}$, *where* $\sum_{i=1}^{k} n_i \theta_i = 0$. *Let* $N \to \infty$, $n_i/N \to \lambda_i$, $0 < \lambda_i < 1$, $i = 1, \ldots, k$, $\sum_{i=1}^{k} \lambda_i = 1$, *and suppose that the following exists*:

$$\lim_{N \to \infty} \int_{-\infty}^{\infty} \sqrt{N} \left[G\left(x - \frac{\theta}{\sqrt{N}} \right) - G(x) \right] dG(x) = -\theta \int_{-\infty}^{\infty} G'(x)\, dG(x).$$

Then H has a limiting noncentral chi-square distribution of $(k-1)$ degrees of freedom and a noncentral parameter

$$\delta = 12 \left(\int_{-\infty}^{\infty} G'(x)\, dG(x) \right)^2 \sum_{i=1}^{k} \lambda_i (\theta_i - \bar{\theta}_.)^2,$$

where $\bar{\theta}_. = \sum_{i=1}^{k} \lambda_i \theta_i$.

Consistency of the Test. Let

$$p_{ij} = \int_{-\infty}^{\infty} F_i(x)\, dF_j(x).$$

Under the null hypothesis $p_{ij} = \frac{1}{2}$. The test is consistent if

$$\sum_{\substack{t=1 \\ t \neq i}}^{k} \lambda_t (p_{it} - \tfrac{1}{2}) \neq 0,$$

for at least one i in $[1, \ldots, k]$. To achieve this it is sufficient to require that $G(x)$ is also strictly increasing.

Pitman Asymptotic Efficiency. In the model $X_{ij} = \mu + \alpha_i + \varepsilon_{ij}$, we will also require that

$$E[\varepsilon_{ij}] = 0, \qquad \sigma^2(\varepsilon_{ij}) = \sigma^2, \qquad E[\varepsilon_{ij}^4] < \infty.$$

We note that the kurtosis γ_2 may be written as

$$\gamma_2 = (E[\varepsilon_{ij}^4]/\sigma^4) - 3.$$

The Pitman asymptotic efficiency of the H-test relative to the classic one-way analysis of variance F-test is given by

$$e_\infty(H, F) = 12\sigma^2 \left(\int_{-\infty}^{\infty} g^2(x)\, dx \right)^2 \geqslant \frac{108}{125} \simeq 0.864,$$

where $g(x)$ is the density associated with the distribution $G(x)$. This expression coincides with the corresponding two-sample problem for the asymptotic relative efficiency of the Wilcoxon–Mann–Whitney test relative to the F-test. [For $g(x) = (\sqrt{2\pi}\,\sigma)^{-1} \exp[-x^2/2\sigma^2]$, $e_\infty(H, F) = 3/\pi$.]

(ii) *A Multiple Comparison.* By an elementary use of the Cauchy–Schwarz inequality one arrives to the conclusion that the $k(k-1)/2$ inequalities

$$|\bar{R}_{i.} - \bar{R}_{j.}| \leqslant \sqrt{\frac{1}{n_i} + \frac{1}{n_j}} \sqrt{\frac{N(N+1)}{12}} (C_N(\alpha))^{1/2} \qquad (*)$$

hold simultaneously with a probability not smaller than $1 - \alpha$, in the notation in (i), and where $P[H \leqslant C_N(\alpha)] = 1 - \alpha$, with the latter probability evaluated under the null hypothesis that $\alpha_1 = \cdots = \alpha_k$. If four any pair (i, j), $i \neq j$, with values in $[1, \ldots, k]$, $|\bar{R}_{i.} - \bar{R}_{j.}|$ does not satisfy the inequality $(*)$ then we infer that $\alpha_i \neq \alpha_j$. For $n_i = m$, $i = 1, \ldots, k$, $m \to \infty$,

$$\sqrt{12/k(N+1)} \max_{i,j} |\bar{R}_{i.} - \bar{R}_{j.}|$$

has a limiting distribution of the range (§3.42) $\max_{i,j}|Y_i - Y_j| = Y_{(k)} - Y_{(1)}$, where $Y_1, \ldots, Y_k$ are independent standard normal variables, and $Y_{(1)} \leqslant \cdots \leqslant Y_{(k)}$. The density of the distribution of $Y_{(k)} - Y_{(1)}$ is obtained from $f_U(u)$ in §3.42, by setting $a = -\infty$, $b = \infty$, $f(z) = (2\pi)^{-1/2} \exp[-z^2/2]$, and by replacing $F(z)$ by the standard normal distribution, and n by k.

See also the following §§1.8, 1.14, 1.15, 1.16, 1.27, 1.33, 1.36, 1.40, 1.43, 2.21, 2.22, 1.23, 1.29, 2.30, 2.31, 2.32, 3.7, 3.8 and 3.10–3.15.

[Cf., Randles and Wolfe (1979), Hájek and Šidák (1967), Manoukian (1986), Bradley (1968, 1969), Durbin (1973), Kendall (1975), Kendall and Stuart (1979), Lehmann (1975), Pratt and Gibbons (1981), Puri and Sen (1971), Noether (1963), Wheeler (1973), Schmetterer (1974).]

CHAPTER 2

Fundamental Limit Theorems

§2.1. Modes of Convergence of Random Variables

(i) A sequence $\{X_n\}$ of random variables is said to converge in probability to a random variable X if for every $\varepsilon > 0$, $\lim_{n\to\infty} P[|X_n - X| < \varepsilon] = 1$. In particular, a sequence $\{X_n\}$ of random variables is said to converge in probability to a constant c if for every $\varepsilon > 0$, $\lim_{n\to\infty} P[|X_n - c| < \varepsilon] = 1$.

(ii) A sequence $\{X_n\}$ of random variables is said to converge in the rth mean, for $r > 0$, to a random variable X if for every $\varepsilon > 0$, $\lim_{n\to\infty} P[|X_n - X|^r < \varepsilon] = 1$. In particular, for $r = 2$, it is said to converge in mean-square.

(iii) A sequence $\{X_n\}$ of random variables with corresponding distributions $\{F_n\}$ is said to converge in distribution (or law) to a random variable X with distribution $F(x)$ if $\lim_{n\to\infty} F_n(x) = F(x)$ at each continuity point x of F. In particular, if X has a normal distribution $N(\mu, \sigma^2)$ with mean μ and variance σ^2, it is said that X_n has a limiting $N(\mu, \sigma^2)$ distribution. Quite generally, if X has a distribution F, then it is said that X_n has a limiting F-distribution.

(iv) A seqeunce $\{X_n\}$ of random variables is said to converge with probability one (also, sometimes, said to converge almost surely or strongly) to a random variable X if $P[\lim_{n\to\infty} X_n = X] = 1$. The following theorem may be useful in establishing the convergence with probability one of a sequence $\{X_n\}$ of random variables to a random variable X: If for every $\varepsilon > 0$, $\sum_{n=1}^{\infty} P[|X_n - X| \geqslant \varepsilon] < \infty$, then X_n converges with probability one to X.

Convergence with probability one implies convergence in probability, and the latter implies convergence in distribution. Convergence in the rth mean implies convergence in probability.

[Cf., Burrill (1972), Cramér (1974), Fourgeaud and Fuchs (1976), Manoukian (1986), Milton and Tsokos (1976), Roussas (1973), Serfling (1980), Tucker (1967), Wilks (1962).]

§2.2. Slutsky's Theorem

Let $\{X_n\}$ be a sequence of random variables converging in distribution to a random variable X. Let $\{Y_n\}$ be a sequence of random variables converging in probability to a constant c. Then:

(i) $\{X_n + Y_n\}$ converge in distribution to $X + c$.
(ii) $\{X_n Y_n\}$ converges in distribution to Xc.
(iii) For $c \neq 0$, $\{X_n/Y_n\}$ converges in distribution to X/c.

[Cf., Serfling (1980), Manoukian (1986).]

§2.3. Dominated Convergence Theorem

Consider a sequence $\{g_n(x)\}$ of (real or complex) functions such that $\lim_{n\to\infty} g_n(x) = g(x)$ with probability P_X-one and $|g_n(x)| \leqslant G(x)$ for all x and n. If $E[G(X)] < \infty$, then $\lim_{n\to\infty} E[g_n(X)] = E[g(X)]$.

[Cf., Cramér (1974), Manoukian (1986).]

§2.4. Limits and Differentiation Under Expected Values with Respect to a Parameter t

(i) Let $h(x, t)$ be a continuous function with respect to t at $t = t_0$ with probability P_X-one, and if $|h(x, t)| \leqslant h_1(x)$, $E[h_1(X)] < \infty$ for all t in some neighborhood of t_0, then $\lim_{t\to t_0} E[h(X, t)] = E[h(X, t_0)]$, where the expectation is with respect to a random variable X, $t_0 \in [-\infty, \infty]$.

(ii) If for all t in some open interval (t_1, t_2), $(\partial/\partial t)h(x, t)$ exists, with probability P_X-one, and is continuous in t, and

$$|(\partial/\partial t)h(x, t)| \leqslant h_2(x), \qquad E[h_2(X)] < \infty,$$

then

$$(d/dt)E[h(X, t)] = E[(\partial/\partial t)h(X, t)], \qquad t \in (t_1, t_2).$$

[Cf., Cramér (1974).]

§2.5. Helly–Bray Theorem

Let $\{F_n\}$ be a sequence of distributions converging for $n \to \infty$ to a distribution F at each of its continuity points. Suppose $g(x)$ is a continuous function and its absolute value is bounded everywhere, then

$$\lim_{n \to \infty} \int_{-\infty}^{\infty} g(x)\, dF_n(x) = \int_{-\infty}^{\infty} g(x)\, dF(x).$$

[Cf., Burrill (1972), Cramér (1974), Manoukian (1986).]

§2.6. Lévy–Cramér Theorem

A sequence $\{F_n(x)\}$ of distributions converges to a distribution $F(x)$ if and only if the sequence $\{\Phi_n(t)\}$ of their corresponding respective characteristic functions converge to a limit $\Phi(t)$ which is continuous at $t = 0$. Also $\Phi(t)$ is identical to the characteristic function corresponding to $F(x)$. For a multivariate generalization simply replace x by $\mathbf{x}$ and t by $\mathbf{t}$.

[Cf., Wilks (1962), Cramér (1974), Manoukian (1986).]

§2.7. Functions of a Sequence of Random Variables

(i) Let $\{X_n\}$ be a sequence of random variables converging in probability to a random variable X. Let $h(x)$ be a continuous function on the real line, then $\{h(X_n)\}$ converges in probability to $h(X)$.

(ii) Let $\{X_n\}$ be a sequence of random variables converging in probability to a finite constant c. Let $h(x)$ be a continuous function in some neighborhood of c, then $\{h(X_n)\}$ converges in probability to $h(c)$.

(iii) Let $\{X_n\}$ be a sequence of random variables, and μ some finite number, such that $\sqrt{n}(X_n - \mu)$ has, for $n \to \infty$, a limiting $N(0, 1)$ distribution. Let $g(x)$ be any real function having a continuous first derivative near μ and $g'(\mu) > 0$, then $\sqrt{n}(g(X_n) - g(\mu))/g'(\mu)$ has, for $n \to \infty$, a limiting $N(0, 1)$ distribution.

[Cf., Wilks (1962).]

§2.8. Weak Laws of Large Numbers

These refer to convergence in probability of averages of random variables.

(i) (Chebyshev). If $\{X_n\}$ is a sequence of independent random variables with the same mean μ and variance σ^2, assumed finite, then $\bar{X} = \sum_{i=1}^{n} X_i/n$ converges in probability to μ.

(ii) (A Generalization). Let $\{X_n\}$ be a sequence of independent random variables with means $\mu_1, \ldots, \mu_n$ and variances $\sigma_1^2, \ldots, \sigma_n^2$, respectively, assumed finite. If

$$\lim_{n\to\infty} (1/n^2) \sum_{i=1}^{n} \sigma_i^2 = 0,$$

then $[\bar{X} - \mu(n)]$ converges in probability to zero, where

$$\mu(n) = \sum_{i=1}^{n} \mu_i/n.$$

[Cf., Wilks (1982), Burrill (1972), Milton and Tsokos (1976).]

§2.9. Strong Laws of Large Numbers

These refer to convergence with probability one of averages of random variables (Kolmogorov's laws).

(i) Let $\{X_n\}$ be a sequence of independent identically distributed random variables, then $\bar{X} = \sum_{i=1}^{n} X_i/n$ converges with probability one to a finite constant c if and only if $E[|X_1|] < \infty$ and $E[X_1] = c$.

(ii) Let $\{X_n\}$ be a sequence of independent random variables with means $\mu_1, \ldots, \mu_n$ and variances $\sigma_1^2, \ldots, \sigma_n^2$, assumed finite. If

$$\lim_{n\to\infty} \sum_{i=1}^{n} \sigma_i^2/i^2 < \infty,$$

then $[\bar{X} - \mu(n)]$ converges with probability one to zero, where

$$\mu(n) = \sum_{i=1}^{n} \mu_i/n.$$

[Cf., Burrill (1972), Milton and Tsokos (1976), Wilks (1962), Tucker (1967).]

§2.10. Berry–Esséen Inequality

Let $X_1, \ldots, X_n$ be independent identically distributed random variables, each with mean μ and variance $\sigma^2 > 0$, and suppose $E[|X_1 - \mu|^3] < \infty$. Let $G_n(x)$ denote the distribution of $\sqrt{n}(\bar{X} - \mu)/\sigma$, where $\bar{X} = \sum_{i=1}^{n} X_i/n$, and let $\phi(x)$ denote the standard normal distribution $\int_{-\infty}^{x} e^{-y^2/2}\, dy/\sqrt{2\pi}$. Then

$$\sup_x |G_n(x) - \phi(x)| \leqslant \frac{A}{\sqrt{n}} \frac{E[|X_1 - \mu|^3]}{\sigma^3}, \qquad n \geqslant 1,$$

where A is a fundamental constant independent of n and is actually not greater than 0.7975 (van Beeck). Also since there is no restriction on the class of distributions of the X_i, with the exception that $E[|X_1 - \mu|^3] < \infty$, a lower bound to the constant A has been obtained (Esséen): A cannot be smaller than $1/\sqrt{2\pi}$. In particular, we note that the right-hand side of the above inequality vanishes for $n \to \infty$.

[Cf., Esséen (1956), van Beeck (1972), Petrov (1975).]

§2.11. de Moivre–Laplace Theorem

If X_n has a binomial distribution with parameters n and p with $0 < p < 1$, then $(X_n - np)/\sqrt{np(1-p)}$ has, for $n \to \infty$, a limiting $N(0, 1)$ distribution.

[Cf., Cramér (1974).]

§2.12. Lindeberg–Lévy Theorem

Let $\{X_n\}$ be a sequence of independent identically distributed random variables each with finite variance $\sigma^2 > 0$ and mean μ. Then $\sqrt{n}(\bar{X} - \mu)/\sigma$ has, for $n \to \infty$, a limiting $N(0, 1)$ distribution, where $\bar{X} = \sum_{i=1}^{n} X_i/n$.

[Cf., Cramér (1974).]

§2.13. Liapounov Theorem

Let $\{X_n\}$ be a sequence of independent, through not necessarily identically distributed, random variables with $E[X_i] = \mu_i$, $E[(X_i - \mu_i)^2] = \sigma_i^2$, and suppose that $\mu_3^{(i)} = E[|X_i - \mu_i|^3] < \infty$ for $i = 1, \ldots, n$. Define

$$\sigma^2(n) = \sum_{i=1}^{n} \sigma_i^2, \qquad \rho^3(n) = \sum_{i=1}^{n} \mu_3^{(i)}, \quad \text{and} \quad \alpha(n) = \sum_{i=1}^{n} \mu_i/n.$$

If $\lim_{n\to\infty} \rho(n)/\sigma(n) = 0$, then $n(\bar{X} - \alpha(n))/\sigma(n)$ has, for $n \to \infty$, a limiting $N(0, 1)$ distribution.

[Cf., Cramér (1974).]

§2.14. Kendall–Rao Theorem

Consider a sequence $X_1, X_2, \ldots$ of random variables with finite rth moments $\alpha_r(1) = E[X_1^r]$, $\alpha_r(2) = E[X_2^r]$, $\ldots$ for all $r = 1, 2, \ldots$, respectively, and suppose that $\lim_{n\to\infty} \alpha_r(n) = \alpha_r$ exist for all $r = 1, 2, \ldots$. If $X_1, X_2, \ldots$ converge in distribution to a random variable X with distribution $F(x)$, then $\alpha_1, \alpha_2, \ldots$ is the moment sequence of $F(x)$. Conversely, if $\alpha_1, \alpha_2, \ldots$ determine, uniquely, a distribution $F(x)$, then $F(x)$ is the limiting distribution of the sequence $X_1, X_2, \ldots$.

[Cf., Wilks (1962), Manoukian (1986).]

§2.15. Limit Theorems for Moments and Functions of Moments

(i) Let $X_1, \ldots, X_n$ be independent identically distributed random variables. Define $\alpha_s = E[X_1^s]$, $s = 1, 2, \ldots$; $\mu_s = E[(X_1 - \alpha_1)^s]$, $s = 2, \ldots$. Also define

$$a_s = \sum_{i=1}^{n} X_i^s/n, \qquad s = 1, 2, \ldots;$$

$$m_s = \sum_{i=1}^{n} (X_i - \bar{X})^s/n, \qquad s = 2, \ldots;$$

$$m_s' = \sum_{i=1}^{n} (X_i - \alpha_1)^s/n, \qquad s = 1, 2, \ldots.$$

Suppose for some s, $\alpha_{2s} < \infty$, then

$$\sqrt{n}(a_s - \alpha_s); \qquad \sqrt{n}(m_s - \mu_s) \quad \text{for } s \neq 1;$$

$\sqrt{n}(m_s' - \mu_s)$, where for $s = 1$ replace μ_s by 0 in the latter, have for $n \to \infty$, limiting normal distributions with zero means, and with variances

$$(\alpha_{2s} - \alpha_s^2);$$

$$(\mu_{2s} - \mu_s^2 - 2s\mu_{s-1}\mu_{s+1} + s^2\mu_2\mu_{s-1}^2);$$

$$(\mu_{2s} - \mu_s^2);$$

respectively, provided the latter do not vanish.

(ii) Let $X_1, \ldots, X_n$ be independent identically distribution random variables each with variance σ^2, $0 < \sigma^2 < \infty$, and kurtosis γ_2, $0 < \gamma_2 + 2 < \infty$. Define the simple variance

$$S^2 = \sum_{i=1}^{n} (X_i - \bar{X})^2/(n - 1).$$

Then, in particular,

$$[\sqrt{n}(S^2 - \sigma^2)/\sigma^2\sqrt{\gamma_2 + 2}], \qquad [2\sqrt{n}(S - \sigma)/\sigma\sqrt{\gamma_2 + 2}],$$

$$[2\sqrt{n}(1 - (\sigma/S))/\sqrt{\gamma_2 + 2}], \qquad [\sqrt{n}\ln(S^2/\sigma^2)/\sqrt{\gamma_2 + 2}],$$

have each, for $n \to \infty$, a limiting $N(0, 1)$ distribution.

(iii) Let $(X_1, \ldots, X_{n_1})$ and $(Y_1, \ldots, Y_{n_2})$ be two independent sets of independent random variables, such that the X_i have a common distribution with variance σ_1^2, $0 < \sigma_1^2 < \infty$, and kurtosis $\gamma_2^{(1)}$, $0 < 2 + \gamma_2^{(1)} < \infty$, and the Y_i have a common distribution with variance σ_2^2, $0 < \sigma_2^2 < \infty$, and kurtosis $\gamma_2^{(2)}$, $0 < 2 + \gamma_2^{(2)} < \infty$. Define the corresponding sample variances S_1^2, S_2^2 as in (ii). Then for $n_1, n_2 \to \infty$, $(n_1/(n_1 + n_2)) \to \lambda$, $0 < \lambda < 1$,

$$\sqrt{n_1 n_2/(n_1 + n_2)}\,[(S_1^2/\sigma_1^2) - (S_2^2/\sigma_2^2)]$$

has a limiting $N(0, v)$ distribution with

$$v = (\sigma_1^4/\sigma_2^4)[(1 - \lambda)(\gamma_2^{(1)} + 2) + \lambda(\gamma_2^{(2)} + 2)].$$

(iv) Let $(X_1, Y_1)', \ldots, (X_n, Y_n)'$ be independent identically distributed random two-vectors. Let

$$E[X_1] = \mu_X, \qquad E[Y_1] = \mu_Y, \qquad \sigma^2(X_1) = \sigma_X^2, \qquad \sigma^2(Y_1) = \sigma_Y^2.$$

$$\rho = E[(X_1 - \mu_X)(Y_1 - \mu_Y)]/\sigma_X\sigma_Y,$$

and set

$$Z = (X_1 - \mu_X)/\sigma_X, \qquad W = (Y_1 - \mu_Y)/\sigma_Y.$$

Define the sample correlation coefficient

$$R = \frac{\sum_{i=1}^{n} (X_i - \bar{X})(Y_i - \bar{Y})}{\sqrt{\sum_{i=1}^{n} (X_i - \bar{X})^2 \sum_{j=1}^{n} (Y_i - \bar{Y})^2}}.$$

Then for $n \to \infty$, $\sqrt{n}(R - \rho)$ has a limiting $N(0, v)$ distribution with

$$v = E[Z^2W^2] - \rho E[Z^3W] - \rho E[ZW^3] - (\rho^2/4)E[(Z^2 + W^2)^2],$$

assuming all these moments exists. For sampling from a bivariate normal distribution the expression for v simplifies to

$$v = (1 - \rho^2)^2.$$

(v) Let $(X_{11}, \ldots, X_{1n_1}), \ldots, (X_{k1}, \ldots, X_{kn_k})$ be k independent sets of independent random variables such that within each group all the variables are identically distributed. Define the corresponding sample variances $S_1^2, \ldots, S_k^2$ based on the sample sizes $n_1, \ldots, n_k$ respectively. Let $\sigma^2(X_{11}) = \sigma_1^2, \ldots, \sigma^2(X_{k1}) = \sigma_k^2$, and denote the corresponding kurtoses by $\gamma_2^{(1)}, \ldots, \gamma_2^{(k)}$. We define the Bartlett statistic:

$$M = -\sum_{j=1}^{k} \nu_j \ln S_j^2 + \nu \ln\left(\sum_{j=1}^{k} \frac{\nu_j}{\nu} S_j^2\right),$$

where $\nu_j = n_j - 1$, $\nu = \sum_{j=1}^{k} \nu_j$. Suppose

$$\gamma_2^{(1)} = \cdots = \gamma_2^{(k)} \equiv \gamma_2, \qquad 0 < 2 + \gamma_2 < \infty,$$

$$\sigma_i = \sigma(1 + (\theta_i/\sqrt{n})),$$

where $n = \sum_{i=1}^{k} n_i$ and σ is some finite positive number. Then for $n_1, \ldots, n_k \to \infty$, such that $\nu_j/\nu \to \lambda_j$, $0 < \lambda_j < 1$, for $j = 1, \ldots, k$, M converges in distribution to $(1 + (\gamma_2/2))\chi_{k-1}^2(\delta)$, where $\chi_{k-1}^2(\delta)$ is a random variable having a noncentral chi-square distribution of $(k - 1)$ degrees of freedom and a noncentrality parameter

$$\delta = [4/(2 + \gamma_2)] \sum_{j=1}^{k} \lambda_j(\theta_j - \bar{\theta}.)^2,$$

where

$$\bar{\theta}. = \sum_{j=1}^{k} \lambda_j \theta_j.$$

In particular, if $\theta_1 = \cdots = \theta_k = 0$, that is, $\sigma_1^2 = \cdots = \sigma_k^2$, then $\delta \equiv 0$, and $\chi_{k-1}^2(0) \equiv \chi_{k-1}^2$ has a chi-square distribution of $(k - 1)$ degrees of freedom.

See also §§3.47, 3.48 and 3.50.

[Cf., Cramér (1974), Serfling (1980), Manoukian (1982, 1984a, 1986), Bartlett (1937).]

§2.16. Edgeworth Expansions

(i) Let $X_1, \ldots, X_n$ be independent identically distributed random variables with $E[X_1] = 0$, $E[X_1^2] = 1$, and suppose $E[X_1^5] < \infty$. Let $F_n(x)$ denote the distribution of $\bar{X} = \sum_{i=1}^n X_i/n$. Then

$$F_n(x) = \phi(x) - \frac{e^{-x^2/2}}{\sqrt{2\pi}} \left\{ \frac{\gamma_1(x^2 - 1)}{6n^{1/2}} + \frac{\gamma_1^2(x^5 - 10x^3 + 15x)}{72n} + \frac{\gamma_2(x^3 - 3x)}{24n} \right\}$$

$$+ R_n(x), \qquad n \geqslant 1,$$

where $\phi(x)$ is the standard normal distribution:

$$\int_{-\infty}^{x} e^{-y^2/2}\, dy/\sqrt{2\pi}, \quad \text{and} \quad \gamma_1 = E[X_1^3], \qquad \gamma_2 = E[X_1^4] - 3,$$

denote the coefficient of skewness and kurtosis of X_1, respectively. Also [e.g., Petrov (1975)] for al x:

$$|R_n(x)| \leqslant c\left[\frac{E[X^4]}{n^2} + \frac{E[|X|^5]}{n^{3/2}} + n^{10}\left(\frac{1}{2n} + \sup_{|t| \geqslant \delta} |\Phi(t)|\right)^n\right], \qquad n \geqslant 1,$$

where $\Phi(t)$ is the characteristic function of X_1, and $\delta = 1/(12E[|X_1|^3])$. We note, in particular, if for any $\varepsilon > 0$, $\sup_{|t| \geqslant \varepsilon} |\Phi(t)| < 1$, then for $n \to \infty$ $((1/2n) + \sup_{|t| \geqslant \delta} |\Phi(t)|)^n$ vanishes faster than any inverse power of n; that is, $R_n(x) = O(n^{-3/2})$.

(ii) For $n = 1$, in particular, let $F(x)$ denote the distribution of X_1, then

$$F(x) = \phi(x) - \frac{e^{-x^2/2}}{\sqrt{2\pi}} \left\{ \frac{\gamma_1(x^2 - 1)}{6} + \frac{\gamma_1^2(x^5 - 10x^3 + 15)}{72} + \frac{\gamma_2(x^3 - 3x)}{24} \right\}$$

$$+ R(x),$$

$$|R(x)| \leqslant c\left\{E[X^4] + E[|X|^5] + \frac{1}{2} + \sup_{|t| \geqslant \delta} |\Phi(t)|\right\}.$$

[Cf., Petrov (1975).]

§2.17. Quantiles

(i) Let $X_1, \ldots, X_n$ denote independent identically distributed random variables each with a continuous distribution $F(x)$. Let $X_{(1)} \leqslant \cdots \leqslant X_{(n)}$ denote the order statistics of $X_1, \ldots, X_n$. Let x_ξ denote an ξth quantile of F. Suppose that x_ξ is unique, $0 < \xi < 1$, $(k/n) \to \xi$, $\xi \leqslant (k/n) \leqslant \xi + (1/n)$ for $n \to \infty$, $k \to \infty$, then in the latter limit $X_{(k)}$ converges in probability to x_ξ.

(ii) If in addition to the conditions given in (i), F has a continuous first derivative f in the neighborhood of x_ξ and $f(x_\xi) > 0$, then

$$[\sqrt{n}f(x_\xi)(X_{(k)} - x_\xi)/\sqrt{\xi(1 - \xi)}],$$

for $n \to \infty$, has a limiting $N(0, 1)$ distribution.

[Note that the differentiability of $F(x)$ guarantees the existence of a unique quantile.]

[Cf., Wilks (1962), David (1981).]

§2.18. Probability Integral Transform with Unknown Location and/or Scale Parameters

Let $X_1, \ldots, X_n$ be independent identically distributed random variables each with a *continuous* distribution $F(x; \theta_1, \theta_2)$ depending only on two (finite) parameters θ_1 and θ_2, such that $F(x; \theta_1, \theta_2) = G((x - \theta_1)/\theta_2)$, and $G(x)$ is independent of θ_1, θ_2. The parameters θ_1 and θ_2 are, respectively, location and scale parameters. If θ_1 and/or θ_2 are unknown, then let $\hat{\theta}_1 = \hat{\theta}_1(X_1, \ldots, X_n)$ and $\hat{\theta}_2 = \hat{\theta}_2(X_1, \ldots, X_n)$ be two statistics of measures of location and scale, respectively, that is for all real λ,

$$\hat{\theta}_1(x_1 + \lambda, \ldots, x_n + \lambda) = \hat{\theta}_1(x_1, \ldots, x_n) + \lambda,$$

$$\hat{\theta}_1(\lambda x_1, \ldots, \lambda x_n) = \lambda\hat{\theta}_1(x_1, \ldots, x_n),$$

$$\hat{\theta}_2(x_1 + \lambda, \ldots, x_n + \lambda) = \hat{\theta}_2(x_1, \ldots, x_n),$$

$$\hat{\theta}_2(\lambda x_1, \ldots, \lambda x_n) = |\lambda|\hat{\theta}_2(x_1, \ldots, x_n).$$

Then (David–Johnson) (§1.8) for all $i = 1, \ldots, n$, the distributions of $G((X_i - \hat{\theta}_1)/\theta_2)$, ($\theta_2$ = known); $G((X_i - \theta_1)/\hat{\theta}_2)$, ($\theta_1$ = known); $G((X_i - \hat{\theta}_1)/\hat{\theta}_2)$ are, *independent* of the parameters θ_1 and θ_2, and hence the latter are statistics.

Suppose in the sequel that $G(x)$ has a continuous and bounded density $g(x)$: $|g(x)| \leqslant c$ for all x.

(i) If θ_1 is unknown, θ_2 is known and $E[|\hat{\theta}_1 - \theta_1|] \to 0$ for $n \to \infty$, then for all $i = 1, \ldots, n$, $G((X_i - \hat{\theta}_1)/\theta_2)$ has, for $n \to \infty$, a limiting uniform distribution on (0, 1), that is, it is asymptotically distribution-free.

(ii) If θ_1 is known, θ_2 is unknown,

$$E[(X_i - \theta_1)^2] < \infty, \qquad E[|\hat{\theta}_2^{-1} - \theta_2^{-1}|^2] \to 0 \quad \text{for } n \to \infty,$$

then for all $i = 1, \ldots, n$, $G((X_i - \theta_1)/\hat{\theta}_2)$ has, for $n \to \infty$, a limiting uniform distribution on (0, 1), that is, it is asymptotically distribution-free.

(iii) If θ_1 and θ_2 are unknown,

$$E[(X_i - \theta_1)^2] < \infty, \qquad E[(\hat{\theta}_1 - \theta_1)^2] \to 0, \qquad E[(\hat{\theta}_2^{-1} - \theta_2^{-1})^2] \to 0,$$
$$\text{for } n \to \infty,$$

then for all $i = 1, \ldots, n$, $G((X_i - \hat{\theta}_1)/\hat{\theta}_2)$ has, for $n \to \infty$, a limiting uniform distribution on (0, 1), that is, it is asymptotically distribution-free.

(iv) In particular, if θ_1 and θ_2 are, respectively, the mean μ and the standard deviation σ, and the latter are estimated, respectively, by the sample mean $\bar{X}$ and the sample standard deviation S, then it is sufficient to assume that the kurtosis γ_2 of G is finite: $0 < \gamma_2 + 2 < \infty$, for the validity of the assumptions in the above theorems.

See also §§1.8 and 1.44.

[Cf., David and Johnson (1948), Manoukian (1984b, 1986).]

§2.19. α-Trimmed Mean

(i) (Bickel). Let $X_1, \ldots, X_n$ be independent identically distributed random variables each with a continuous, symmetric (i.e., $F(x) = 1 - F(-x)$) and strictly increasing distribution $F(x)$ and probability density $f(x) \geqslant 0$. Suppose $f(x)$ is continuous and nonvanishing for $a < x < b$ (a and b may be $-\infty$ and $+\infty$). Define the α-trimmed mean of the sample:

$$\bar{X}_\alpha = (n - 2[n\alpha])^{-1} \sum_{i=[n\alpha]+1}^{n-[n\alpha]} X_{(i)}, \qquad 0 < \alpha < \tfrac{1}{2},$$

where $X_{(1)} \leqslant \cdots \leqslant X_{(n)}$ are the other statistics, and $[x]$ is the largest integer $\leqslant x$. Then, for $n \to \infty$, $\sqrt{n}\,\bar{X}_\alpha$ has a limiting $N(0, \sigma_\alpha^2)$ distribution, where

$$\sigma_\alpha^2 = 2(1 - 2\alpha)^{-2}\left[\int_0^{x_{1-\alpha}} x^2 f(x)\,dx + \alpha x_{1-\alpha}^2\right],$$

provided $\sigma_\alpha > 0$ and where $x_{1-\alpha}$ is the (unique)$(1 - \alpha)$th quantile of $F(x)$, that is $F(x_{1-\alpha}) = 1 - \alpha$.

(ii) (Bickel–Jaeckel). Define the Winsorized sample variance:

$$\hat{\sigma}_\alpha^2 = \frac{(1 - 2\alpha)^{-2}}{n}\left\{\sum_{i=[n\alpha]+1}^{n-[n\alpha]} (X_{(i)} - \bar{X}_\alpha)^2 + [n\alpha] \right.$$
$$\left. \times [(X_{([n\alpha]+1)} - \bar{X}_\alpha)^2 + (X_{(n-[n\alpha])} - \bar{X}_\alpha)^2]\right\}.$$

Suppose $E[X_1^2] < \infty$, then under the conditions stated in (i), $\hat{\sigma}_\alpha^2$ converges, for $n \to \infty$, in probability to σ_α^2, where σ_α^2 is defined in (i). Therefore, we also have, that $\sqrt{n}\,\bar{X}_\alpha/\hat{\sigma}_\alpha$ has, for $n \to \infty$, a limiting $N(0, 1)$ distribution.

(iii) (Bickel). Suppose $E[X_1^2] < \infty$, then under the conditions given in (i), the asymptotic efficiency of $\bar{X}_\alpha$ to $\bar{X}$ is given by

$$e_\infty(\bar{X}_\alpha, \bar{X}) = \left(\int_{-\infty}^{\infty} x^2 f(x)\,dx\right)\Big/\sigma_\alpha^2 \geqslant (1 - 2\alpha)^2.$$

See also §1.42.

[Cf., Bickel (1965), Jaeckel (1971). See also Serfling (1980).]

§2.20. Borel's Theorem

Let $X_1, \ldots, X_n$ be independent identically distributed random variables each with distribution $F(x)$. Define the sample distribution $\hat{F}_n(x) = [\text{number of } X_i \leqslant x]/n$. Then, for $n \to \infty$, $\hat{F}_n(x)$ converges with probability one to $F(x)$.

[Cf., Burrill (1972), Manoukian (1986).]

§2.21. Glivenko–Cantelli Theorem

Let $X_1, \ldots, X_n$ denote independent identically distributed random variables each with distribution $F(x)$. Define the sample distribution $\hat{F}_n(x) = [\text{number of } X_i \leqslant x]/n$. Then, for $n \to \infty$, $\sup_x |\hat{F}_n(x) - F(x)|$ converges with probability one to zero.

[Cf., Gnedenko (1966), Manoukian (1986).]

§2.22. Kolmogorov–Smirnov Limit Theorems

(i) Let $X_1, \ldots, X_n$ be independent identically distributed random variables each with a *continuous* distribution $F(x)$. Let $\hat{F}_n(x)$ denote the sample (empirical) distribution $\hat{F}_n(x) = [\text{number of } X_i \leqslant x]/n$. Define the statistics:

$$D_n = \sup_x |\hat{F}_n(x) - F(x)|,$$

$$D_n^+ = \sup_x (\hat{F}_n(x) - F(x)),$$

$$D_n^- = \sup_x (F(x) - \hat{F}_n(x)).$$

Then for $n \to \infty$,

$$P[\sqrt{n}\,D_n \leqslant x] \to \left[1 - 2\sum_{k=1}^{\infty} (-1)^{k-1} \exp(-2k^2x^2)\right],$$

$$P[\sqrt{n}\, D_n^+ \leqslant x] \to 1 - \exp(-2x^2),$$
$$P[\sqrt{n}\, D_n^- \leqslant x] \to 1 - \exp(-2x^2), \qquad \text{where } x > 0.$$

(ii) Let $X_1, \ldots, X_{n_1}$; $Y_1, \ldots, Y_{n_2}$ be independent identically distributed random variables each with a *continuous* distribution $F(x)$. Let

$$\hat{F}_{n_1}(x) = \frac{[\text{number of } X_i \leqslant x]}{n_1},$$

$$\hat{G}_{n_2}(x) = \frac{[\text{number of } Y_i \leqslant x]}{n_2}.$$

Define the statistics

$$D_{n_1,n_2} = \sup_x |\hat{F}_{n_1}(x) - \hat{G}_{n_2}(x)|,$$

$$D_{n_1,n_2}^+ = \sup_x (\hat{F}_{n_1}(x) - \hat{G}_{n_2}(x)),$$

and set $N = n_1 n_2/(n_1 + n_2)$. Then for $n_1, n_2 \to \infty$, $(n_1/n_2) \to \lambda > 0$,

$$P[\sqrt{N}\, D_{n_1,n_2} \leqslant x] \to \left[1 - 2 \sum_{k=1}^{\infty} (-1)^{k-1} \exp(-2k^2x^2)\right],$$

$$P[\sqrt{N}\, D_{n_1,n_2}^+ \leqslant x] \to [1 - \exp(-2x^2)], \quad \text{where } x > 0.$$

See also §§1.44, 3.14 and 3.15.

[Cf., Wilks (1962), Hájek and Šidák (1967), Durbin (1973), Manoukian (1986).]

§2.23. Chi-Square Test of Fit

(i) Let $X_1, \ldots, X_n$ be independent identically distributed random variables. We group the observations into k mutually exclusive classes (intervals) $c_1, \ldots, c_k$. Let p_i denote the probability that an observation falls in the ith class c_i. Let N_i be the random variable representing the number of observations falling in the ith class, $\sum_{i=1}^k N_i = n$.

As a test of fit, the hypothesis is that the data $X_1, \ldots, X_n$ may be adjusted by a theoretical distribution (with associated probabilities $p_1, \ldots, p_k$). A conservative recommendation on the choice of the number of classes is to have $np_i \geqslant 5$ for each i. Recent work (Yarnold) suggests that if the number of classes $k \geqslant 3$, and if r denotes the number of expectations $np_i < 5$, then the minimum expectation may be as small as $5r/k$. The hypothesis is rejected for large values of the statistic $\sum_{i=1}^k (N_i - np_i)^2/np_i$.

A comparison of this test with the Kolmogorov–Smirnov test is made in §1.44.

In general, the probabilities p_i depend on n: $p_i = p_i(n)$. If

$$\lim_{n\to\infty} p_i(n) = \hat{p}_i, \qquad 0 < \hat{p}_i < 1 \quad \text{for all } i = 1, \ldots, k \qquad \left(\sum_{i=1}^{k} \hat{p}_i = 1\right),$$

then the statistic

$$\sum_{i=1}^{k} [(N_i - np_i(n))^2/np_i(n)]$$

has, for $n \to \infty$, a limiting chi-square distribution of $(k - 1)$ degrees of freedom.

(ii) Suppose that the probabilities that the observations fall in classes $c_1, \ldots, c_k$ are not necessarily $p_1(n), \ldots, p_k(n)$, but are given by

$$p_i'(n) = p_i(n) + (K_i/\sqrt{n}), \qquad i = 1, \ldots, n,$$

where $K_1, \ldots, K_k$ are some finite numbers and $\sum_{i=1}^{k} K_i = 0$.
If for $n \to \infty$

$$(p_i(n) \to \hat{p}_i), \qquad p_i'(n) \to \hat{p}_i, \qquad 0 < \hat{p}_i < 1, \qquad i = 1, \ldots, k,$$

then

$$\sum_{i=1}^{k} [(N_i - np_i(n))^2/np_i(n)]$$

has a limiting noncentral chi-square distribution of $(k - 1)$ degrees of freedom and a noncentrality parameter

$$\delta = \sum_{i=1}^{k} K_i^2/\hat{p}_i.$$

(iii) (Yarnold). As in (i) except suppose that

$$\lim_{n\to\infty} p_i(n) = 0, \qquad \lim_{n\to\infty} np_i(n) = m_i, \qquad 0 < m_i < \infty \quad \text{for } i = 1, \ldots, r;$$

and

$$\lim_{n\to\infty} p_j(n) = \hat{p}_j, \qquad 0 < \hat{p}_j < 1 \quad \text{for } j = r + 1, \ldots, k.$$

We note that $\sum_{j=r+1}^{k} \hat{p}_j = 1$. Then

$$\sum_{i=1}^{k} [(N_i - np_i(n))^2/np_i(n)],$$

for $n \to \infty$, converges in distribution to the sum $V_1 + V_2$ of two *independent* random variables, where V_1 has a chi-square distribution of $(k - r)$ degrees of freedom, and

$$V_2 = \sum_{i=1}^{k} (U_i - m_i)^2/m_i,$$

where $U_1, \ldots, U_r$ are independent random variables having Poisson distributions with means $m_1, \ldots, m_r$, respectively.

(iv) (Cramér). As in (i), except suppose that the p_i depend on s ($<k-1$) unknown parameters $\theta_1, \ldots, \theta_s$:

$$p_i = p_i(\theta_1, \ldots, \theta_s) \equiv p_i(\boldsymbol{\theta}),$$

and these parameters are estimated from the equations

$$\sum_{i=1}^{k} N_i[(\partial/\partial\theta_j) \ln p_i(\boldsymbol{\theta})] = 0, \qquad j = 1, \ldots, s.$$

Let $\hat{\theta}_{1n}, \ldots, \hat{\theta}_{sn}$ denote corresponding solutions, and write $(\hat{\theta}_{1n}, \ldots, \hat{\theta}_{sn})' \equiv \hat{\boldsymbol{\theta}}_n$. Then

$$\sum_{i=1}^{k} [(N_i - np_i(\hat{\boldsymbol{\theta}}_n))^2/np_i(\hat{\boldsymbol{\theta}}_n)]$$

has, for $n \to \infty$, a limiting chi-square distribution of $(k-s-1)$ degrees of freedom. We note that the unknown parameters $\theta_1, \ldots, \theta_s$ here are estimated from the *grouped* data rather than from the original data. [In practice, these parameters are estimated, however, by the method of moments of estimation (from the *grouped* data).]

See also §1.44(A)(v).

[Darling (1957), Slaketer (1965, 1966), Cramér (1974), Manoukian (1986), Yarnold (1970), Cochran (1952), Hogg (1978).]

§2.24. Maximum Likelihood Estimators

(i) Let $X_1, \ldots, X_n$ denote independent identically distributed random variables each with distribution F_{θ_0} depending on some parameter θ_0. Suppose F_{θ_0} has a probability density or a probability mass function $f(x; \theta_0)$. The likelihood function is defined by

$$L(\theta_0; X_1, \ldots, X_n) = \prod_{i=1}^{n} f(X_i; \theta_0).$$

Suppose that for all θ in some closed interval I containing θ_0:

(a) $$(\partial/\partial\theta) \ln f(x; \theta), \qquad (\partial/\partial\theta)^2 \ln f(x; \theta), \qquad (\partial/\partial\theta)^3 \ln f(x; \theta)$$

exist for all x,

(b) $$|(\partial/\partial\theta) f(x; \theta)| \leqslant g_1(x), \qquad |(\partial/\partial\theta)^2 f(x; \theta)| \leqslant g_2(x),$$
$$|(\partial/\partial)^3 \ln f(x; \theta)| \leqslant g_3(x)$$

hold for all x, and

$$\int_{-\infty}^{\infty} g_1(x)\, dx < \infty, \qquad \int_{-\infty}^{\infty} g_2(x)\, dx < \infty, \qquad E_{\theta_0}[g_3(X)] < \infty.$$

[For discrete variables, replace the integrals signs by summations.]

(c) $$0 < E_\theta[((\partial/\partial\theta)\ln f(X;\theta))^2] < \infty.$$

Then the likelihood equation

$$(\partial/\partial\theta)L(\theta; X_1, \ldots, X_n) = 0$$

admits, for $n \to \infty$, a sequence solution $\{\hat{\theta}_n\}$ converging in probability to θ_0, and

$$\sqrt{n}(\hat{\theta}_n - \theta_0)/\sqrt{E_{\theta_0}[((\partial/\partial\theta_0)\ln f(X;\theta_0))^2]}$$

has a limiting $N(0, 1)$ distribution. The latter, in particular, implies that $\hat{\theta}_n$ is an asymptotic efficient estimate of θ_0. [We note that if $(\partial^2/\partial\theta^2)\ln f(x;\theta)$ is continuous in θ in some neighborhood of θ_0 uniformly in x, and $|\partial^2/\partial\theta^2 \ln f(x;\theta)| \leqslant g_4(x)$ for all x, $E_{\theta_0}[g_4(X)] < \infty$, then the sequence solution $\{\hat{\theta}_n\}$, for $n \to \infty$, does provide a maximum.]

(ii) As in (i) suppose that for all θ in some closed interval containing θ_0:

(a) $$(\partial/\partial\theta)\ln f(x;\theta), \qquad (\partial^2/\partial\theta^2)\ln f(x;\theta)$$

exist and are continuous in θ for all x.

(b) $$|(\partial/\partial\theta)\ln f(x;\theta)| \leqslant g_1(x), \qquad |(\partial^2/\partial\theta^2)\ln f(x;\theta)| \leqslant g_2(x)$$

and

$$E_{\theta_0}[g_1(X)] < \infty, \qquad E_{\theta_0}[g_2(X)] < \infty.$$

(c) $$E_{\theta_0}[\ln f(X,\theta)] \neq E_{\theta_0}[\ln f(X,\theta_0)] \quad \text{for } \theta \neq \theta_0,$$
$$0 < E_{\theta_0}[((\partial/\partial\theta_0)\ln f(X,\theta_0))^2] < \infty.$$

Then the likelihood equation

$$(\partial/\partial\theta)L(\theta; X_1, \ldots, X_n) = 0$$

admits, for $n \to \infty$, a sequence solution $\{\hat{\theta}_n\}$ converging in probability to θ_0, and

$$\sqrt{n}(\hat{\theta}_n - \theta_0)/\sqrt{E_{\theta_0}[((\partial/\partial\theta_0)\ln f(X;\theta_0))^2]}$$

has a limiting $N(0, 1)$ distribution.

[Cf., Fourgeaud and Fuchs (1967), Serfling (1980), Cramér (1974).]

§2.25. M-Estimators

(i) Let $X_1, \ldots, X_n$ be independent identically distributed random variables each with a distribution $F(x)$. Consider a function $\psi(x, t)$, depending on a parameter t, such that $E[\psi(X, t_0)] = 0$ for some $t = t_0$. Suppose that for all t in some closed interval containing t_0:

(a) $(\partial/\partial t)\psi(x, t)$, $(\partial^2/\partial t^2)\psi(x, t)$ exist for all x.

(b) $$|(\partial/\partial t)\psi(x, t)| \leqslant g_1(x), \qquad |(\partial^2/\partial t^2)\psi(x, t)| \leqslant g_2(x)$$

and

$$E[g_1(X)] < \infty, \qquad E[g_2(X)] < \infty.$$

(c) $$0 < |E[(\partial/\partial t_0)\psi(x, t_0)]| < \infty, \qquad 0 < E[\psi^2(X, t_0)] < \infty.$$

Then

$$\sum_{i=1}^{n} \psi(X_i, t) = 0$$

admits, for $n \to \infty$, a sequence solution $\{\hat{T}_n\}$ converging in probability to t_0, and

$$\sqrt{n}(\hat{T}_n - t_0)$$

has a limiting $N(0, \sigma_0^2)$ distribution, where

$$\sigma_0^2 = \frac{E[\psi^2(X, t_0)]}{(E[(\partial/\partial t_0)\psi(X, t_0)])^2}.$$

(ii) As in (i) except instead of the conditions (a)–(c) there suppose that:

(a) $\psi(x, t)$, $(\partial/\partial t)\psi(x, t)$ exist and are continuous at t_0 for all x.

(b) $$|\psi(x, t)| \leqslant g_1(x), \qquad |(\partial/\partial t)\psi(x, t)| \leqslant g_2(x),$$

and

$$E[g_1(X)] < \infty, \qquad E[g_2(X)] < \infty,$$

for all t in some neighborhood of t_0.

(c) $E[\psi(X, t)]$ is strictly increasing (or decreasing) in some neighborhood of t_0.

(d) $$0 < |E[(\partial/\partial t_0)\psi(X, t_0)]| < \infty, \qquad 0 < E[\psi^2(X, t_0)] < \infty.$$

Then

$$\sum_{i=1}^{n} \psi(X_i, t) = 0$$

admits, for $n \to \infty$, a sequence solution $\{\hat{T}_n\}$ converging in probability to t_0, and

$$\sqrt{n}(\hat{T}_n - t_0)$$

has a limiting $N(0, \sigma_0^2)$ distribution, where

$$\sigma_0^2 = \frac{E[\psi^2(X, t_0)]}{(E[(\partial/\partial t_0)\psi(X, t_0)])^2}.$$

(iii) Let $X_1, \ldots, X_n$ be independent identically distributed random variables. Let $\psi(x, t)$ be a function depending on a parameter t, such that $E[\psi(X, t_0)] = 0$ for some $t = t_0$. Suppose that $\psi(x, t)$ is continuous at t_0 for all x, $|\psi(x, t)| \leqslant g(x)$ and $E[g(X)] < \infty$ for all t in some open interval containing t_0. Also suppose that $E[\psi(X, t)]$ is strictly increasing (or decreasing) in the neighborhood of t_0. Then $\sum_{i=1}^{n} \psi(X_i, t) = 0$ admits, for $n \to \infty$, a solution $\hat{T}_n$ converging in probability one to t_0. [Note that the conditions imposed on $\psi(x, t)$ here, guarantees, self consistently, of the existence of a root $t = t_0$: $E[\psi(X, t_0)] = 0$.]

[Cf., Serfling (1980), Fourgeaud and Fuchs (1967), Huber (1977), Lehmann (1983).]

§2.26. Likelihood Ratio Statistic

Under the stated conditions of the existence and asymptotic normality, in distribution, of a maximum likelihood estimator $\hat{\theta}_n$ of the likelihood equation (§2.24) $(\partial/\partial\theta)L(\theta; X_1, \ldots, X_n) = 0$, set

$$\Lambda_n = L(\theta_0; X_1, \ldots, X_n)/L(\hat{\theta}_n; X_1, \ldots, X_n),$$

then, for $n \to \infty$, $-2 \ln \Lambda_n$ has a limiting chi-square distribution of one degree of freedom when θ_0 is the true parameter, and where we recall then $\{\hat{\theta}_n\}$ converges in probability to θ_0. If the conditions stated in (ii) of §2.24 on the maximum likelihood estimators are satisfied, then $P_\theta[-2 \ln \lambda_n > c_n] \to 1$ for $n \to \infty$, $\theta \neq \theta_0$, where

$$P_{\theta_0}[-2 \ln \lambda_n > c_n] \to \alpha, \qquad 0 < \alpha < 1.$$

[This establishes the consistency of the likelihood ratio tests.]

[Cf., Serfling (1980), Wilks (1962), Fourgeaud and Fuchs (1967).]

§2.27. On Some Consistency Problems of Tests

(i) (Lehmann). Consider a test of hypothesis on a parameter θ: H_0: $\theta \in \omega$ against H_A: $\theta \in \Omega - \omega$, where Ω is the parameter space, and suppose that, with a level of significance α, H_0 is rejected for $T_n \geqslant k_n$, where T_n is the test statistic under consideration based on a sample of size n. Suppose that for each $\theta \in \Omega$, T_n converges, for $n \to \infty$, in probability to a number $T(\theta)$, with $T(\theta) \equiv k$ for all $\theta \in \omega$, and $T(\theta) > k$ for all $\theta \in \Omega - \omega$, and $\lim_{n\to\infty} k_n = k_\infty$, $k_\infty \leqslant k$. Then $\{T_n\}$ is a consistent sequence of tests against all alternatives in H_A, that is, $\lim_{n\to\infty} P_\theta[T_n \geqslant k_n] = 1$ for each $\theta \in \Omega - \omega$.

(ii) (Lehmann). Consider a test of hypothesis on a parameter θ: H_0: $\theta \in \omega$ against H_A: $\theta \in \Omega - \omega$, where Ω is the parameter space, and suppose that, with a level of significance α, H_0 is rejected if $T_n \leqslant k_n^{(1)}$ or if $T_n \geqslant k_n^{(2)}$, where T_n is the test statistic under consideration based on a sample of size n. Suppose that for each $\theta \in \Omega$, T_n converges, for $n \to \infty$, in probability to a number $T(\theta)$, with $T(\theta) \equiv k$ for all $\theta \in \omega$, and $T(\theta) \neq k$ for all $\theta \in \Omega - \omega$, and $\lim_{n\to\infty} k_n^{(1)} = k_\infty^{(1)} \geqslant k$, $\lim_{n\to\infty} k_n^{(2)} = k_\infty^{(2)} \leqslant k$. Then $\{T_n\}$ is a consistent sequence of tests against all alternatives in H_A, that is,

$$\lim_{n\to\infty} P_\theta[T_n \leqslant k_n^{(1)} \text{ or } T_n \geqslant k_n^{(2)}] = 1 \quad \text{for each } \theta \in \Omega - \omega.$$

(iii) (Kruskal). Consider a test of hypothesis on a parameter θ (which may be a k-vector): H_0: $\theta \in \omega$ against H_A: $\theta \in \Omega - \omega$, and suppose that, with a level of significance α, H_0 is rejected if $V_n \geqslant t_n^2$, where the statistic V_n in question is of the form $V_n = \sum_{j=1}^k (V_j(n))^2$. Suppose that for each $j = 1, \ldots, k$, the statistic $V_j(n)/\sqrt{n}$ converges, for $n \to \infty$, in probability to a number ξ_j, and that the sequence $\{t_n^2\}$ converges, for $n \to \infty$, to a number $t_0 \geqslant 0$. If for at least one j in $[1, \ldots, k]$, $\xi_j \neq 0$, then $\lim_{n\to\infty} P[V_n \geqslant t_n^2] = 1$.

[Cf., Randles and Wolfe (1979), Kruskal (1952), Manoukian (1986).]

§2.28. Pitman Asymptotic Efficiency

(i) (Noether). Consider a test of hypothesis concerning a parameter θ, H_0: $\theta = \theta_0$ against a sequence of alternatives $\{\theta_n = \theta_0 + (k(n)/\sqrt{n})\}$, where n is the sample size, *and* $k(n) \to k > 0$ for $n \to \infty$. Let T_n be the statistic in question, and let $\psi_n(\theta)$, $\sigma_n(\theta)$ be two functions of θ, which may depend on n. We suppose that the null hypothesis H_0 is rejected for $[T_n - \psi_n(\theta_0)]/\sigma_n(\theta_0) \geqslant a_n$. Suppose

(a) $(\partial/\partial\theta)\psi_n(\theta) = \psi_n'(\theta)$ exists and is continuous in some closed interval containing θ_0, and $\psi_n'(\theta_0) \neq 0$.

(b)
$$\lim_{n\to\infty} \sqrt{n}\,\sigma_n(\theta_0) \neq 0, \qquad \lim_{n\to\infty} [\psi_n'(\theta_0)/\sqrt{n}\,\sigma_n(\theta_0)] > 0,$$
$$\lim_{n\to\infty} [\psi_n'(\theta_n)/\psi_n'(\theta_0)] = 1, \qquad \lim_{n\to\infty} [\sigma_n(\theta_n)/\sigma_n(\theta_0)] = 1.$$

(c) $[T_n(\theta) - \psi_n(\theta)]/\sigma_n(\theta)$, under the null hypothesis, that is, for $\theta = \theta_0$, and under the alternative hypotheses, that is, for $\theta = \theta_n$, has a limiting $N(0, 1)$ distribution, for $n \to \infty$.

The efficacy of the test is *defined* by

$$c = \lim_{n\to\infty} [\psi_n'(\theta_0)/\sqrt{n}\,\sigma_n(\theta_0)].$$

Let $T^{(1)}$ and $T^{(2)}$ be two test statistics with corresponding sample sizes n_1, n_2, and parameters

$$k^{(1)}, k^{(2)}, \qquad \psi_{1n_1}'(\theta_0), \psi_{2n_2}'(\theta_0),$$
$$\sigma_{1n_1}(\theta_0), \sigma_{2n_2}(\theta_0), \qquad c_1, c_2,$$

respectively. The Pitman asymptotic efficiency ($n \to \infty$) of the $T^{(1)}$-test to the $T^{(2)}$-test is

$$e_\infty(T^{(1)}, T^{(2)}) = (c_1/c_2)^2.$$

(ii) In k-sample problems of samples of sizes $n_1, \ldots, n_k$ one may typically consider a test of hypothesis on k parameters $\theta_1, \ldots, \theta_k$: H_0: $\theta_1 = \cdots = \theta_k \equiv 0$, against a sequence of alternatives $\{\theta_i = \alpha_i/\sqrt{N}, i = 1, \ldots, k\}_N$, where not all the α_i are identical, and $N = \sum_{i=1}^k n_i$. The following situation then often arises. The statistic T_N, under consideration for the test, has, for $N \to \infty$, a limiting noncentral chi-square distribution with v degrees of freedom and a noncentrality parameter δ, such that $\delta = 0$ only under the null hypothesis. Let $T_{N_1}^{(1)}$ and $T_{N_2}^{(2)}$ be two test statistics, based on N_1 and N_2 total observations, having limiting noncentral chi-square distribution with v_1 and v_2 degrees of freedom and noncentrality parameters δ_1 and δ_2, respectively, with $v_1 = v_2$. Imposing identical alternative hypotheses $\alpha_i^{(1)}/\sqrt{N_1} = \alpha_i^{(2)}/\sqrt{N_2}$, for $i = 1$, $\ldots, n$, for the two tests, one obtains that the two noncentrality parameters δ_1 and δ_2 are related through an equation: $\delta_1 = (N_1/N_2)c\delta_2$ for N_1, $N_2 \to \infty$. Hence to achieve identical powers for the two tests we must have $c = (N_2/N_1)$ for $N_1, N_2 \to \infty$. Thus the Pitman asymptotic efficiency of the $T^{(1)}$-test to the $T^{(2)}$-test is simply given, in such cases, by $e_\infty(T^{(1)}, T^{(2)}) = c$, and is obtained by equating the noncentrality parameters for N_1, $N_2 \to \infty$. The situation in (i) may be also recast in this form.

[Cf., Noether (1955), Manoukian (1986).]

§2.29. Hodges–Lehmann Estimators

(i) *One-Sample Problem.* Let $X_1, \ldots, X_n$ be independent identically distributed random variables each with a continuous distribution $F(x - \theta)$, where $F(x)$ is symmetric about the origin ($F(x) = 1 - F(-x)$). Consider the test of hypothesis H_0: $\theta = 0$ against H_A: $\theta > 0$, and suppose that H_0 is rejected whenever $T(x_1, \ldots, x_n) \geqslant d_n$, where $T(X_1, \ldots, X_n)$ is the statistic in question. Suppose

(a) There are functions $\psi_n(\theta)$, $\sigma_n(\theta)$ such that the conditions stated in §2.28(i), in the Pitman asymptotic efficiency, in obtaining the efficacy c of the test hold.
(b) The statistic $T(X_1, \ldots, X_n)$ satisfies the two conditions stated in defining the Hodges–Lehmann estimation $\hat{\theta}$, in the one-sample problem (§1.15), with ξ_n denoting the symmetry point.
(c) $$\lim_{n\to\infty} [\xi_n - \psi_n(0)]/\sigma_n(0) = 0.$$

Then $\sqrt{n}\,c(\hat{\theta} - \theta)$ has a limiting $N(0, 1)$ distribution.

For example, for the Wilcoxon signed rank test (see §1.44(A)(ii)), we may take $T = W/(n(n-1)/2)$, $\xi_n = (n+1)/(2(n-1)$ (see §3.7), $\sigma_n(\theta) = 1/\sqrt{3n}$,

$\psi_n(\theta) = E_\theta[T]$, $\psi_n(0) = \xi_n$. Then

$$\xi_n - \psi_n(0) \equiv 0. \qquad \lim_{n\to\infty} ([(\partial/\partial\theta)\psi_n(\theta)]_{\theta=0}) = 2\int_{-\infty}^{\infty} f^2(x)\,dx,$$

$c = \sqrt{12}\int_{-\infty}^{\infty} f^2(x)\,dx > 0$, where $f(x)$ is the probability density associated with $F(x)$.

(ii) *Two-Sample Problem.* Let $X_{11}, \ldots, X_{1n_1}, X_{21}, \ldots, X_{2n_2}$ be two independent samples from continuous distributions $F(x)$ and $F(x - \theta)$, respectively. Consider the test of hypothesis $H_0\colon \theta = 0$ against $H_A\colon \theta > 0$, and suppose that H_0 is rejected whenever $T(x_{11}, \ldots, x_{1n_1}; x_{21}, \ldots, x_{2n_2}) \geqslant t_{n_1,n_2}$, where $T(X_{11}, \ldots, X_{1n_1}; X_{21}, \ldots, X_{2n_2})$ is the statistic in question. Let $n = n_1 + n_2$, consider the limit $n \to \infty$ such that $(n_1/n) \to \lambda$, $0 < \lambda < 1$, $(n_2/n) \to 1 - \lambda$. Suppose

(a) The conditions stated in §2.28(i), in the Pitman asymptotic efficiency, in obtaining the efficacy c, hold.
(b) The statistic $T(X_{11}, \ldots, X_{1n_1}; X_{21}, \ldots, X_{2n_2})$ satisfies the two conditions stated in defining the Hodges–Lehmann estimator $\hat{\theta}$, in the two-sample problem (§1.15), with ξ_{n_1,n_2} denoting the symmetry point.
(c) $$\lim_{n\to\infty} [\xi_{n_1,n_2} - \psi_n(0)]/\sigma_n(0) = 0.$$

Then $\sqrt{n}\,c(\hat{\theta} - \theta)$ has, for $n \to \infty$, a limiting $N(0, 1)$ distribution.

For example, for the Wilcoxon–Mann–Whitney test (see §1.44(B)(i)) we may take $T = U$, where

$$U = W_0 - \frac{n_2(n_2+1)}{2},$$

$$\xi_{n_1,n_2} = \frac{n_1 n_2}{2} \quad \text{(see §3.8)},$$

$$\sigma_n(\theta) = \sqrt{\frac{n_1 n_2 (n_1 + n_2)}{12}},$$

$$n \equiv n_1 + n_2\ (= N), \qquad \psi_n(\theta) = E_\theta[U], \qquad \psi_n(0) = \xi_{n_1,n_2}.$$

Then

$$\xi_{n_1,n_2} - \psi_n(0) \equiv 0, \qquad (\partial/\partial\theta)\psi_n(\theta)|_{\theta=0} = n_1 n_2 \int_{-\infty}^{\infty} f^2(x)\,dx > 0,$$

$$c = \sqrt{12\lambda(1-\lambda)}\int_{-\infty}^{\infty} f^2(x)\,dx > 0,$$

where $f(x)$ is the probability density associated with $F(x)$.

See also §1.44.

[Cf., Randles and Wolfe (1979), Hodges and Lehmann (1963).]

§2.30. Hoeffding's Theorems for U-Statistics

(i) *One-Sample Problems.* In the definition of the one-sample U-statistics (§1.16), if

$$E[h^2(X_1, \ldots, X_m)] < \infty,$$

and

$$0 < \xi_1 = E[h(X_1, \ldots, X_m)h(X_1, X_{m+1}, \ldots, X_{2m-1})] \, (< \infty),$$

then, for $n \to \infty$, $(\sqrt{n/m^2\xi_1})[U - \theta]$ has a limiting $N(0, 1)$ distribution.

(ii) *Two-Sample Problems* (Lehmann). In the definition of the two-sample U-statistics (§1.16), let $n_1 + n_2 = N$, and for $N \to \infty$, let $n_1/N \to \lambda$, $0 < \lambda < 1$. If

$$E[h^2(X_{11}, \ldots, X_{1m_1}; X_{21}, \ldots, X_{2m_2})] < \infty$$

and

$$0 < v \equiv [m_1^2\xi_{1,0}/\lambda] + [m_2^2\xi_{0,1}/(1 - \lambda)] \, (< \infty),$$

then for $N \to \infty$, $(\sqrt{N/v})(U - \theta)$ has a limiting $N(0, 1)$ distribution, where

$$\begin{aligned} \xi_{1,0} &= E[h(X_{11}, \ldots, X_{1m_1}; X_{21}, \ldots, X_{2m_2}) \\ &\quad \times h(X_{11}, X_{1m_1+1}, \ldots, X_{12m_1-1}; X_{2m_2+1}, \ldots, X_{22m_2})] - \theta^2, \\ \xi_{0,1} &= E[h(X_{11}, \ldots, X_{1m_1}; X_{21}, \ldots, X_{2m_2}) \\ &\quad \times h(X_{1m_1+1}, \ldots, X_{12m_2}; X_{21}, X_{2m_2+1}, \ldots, X_{22m_2-1})] - \theta^2. \end{aligned}$$

(iii) *k-Sample Problems* (Lehmann). In the definition of the k-sample U-statistics (§1.16), let $\sum_{i=1}^k n_i = N$, and for $N \to \infty$, let $n_i/N \to \lambda_i$, $0 < \lambda_i < 1$, $i = 1, \ldots, k$, $\sum_{i=1}^k \lambda_i = 1$. If

$$E[h^2(X_{11}, \ldots, X_{1m_1}; \ldots; X_{k1}, \ldots, X_{km_k})] < \infty,$$

and

$$0 < v \equiv \sum_{t=1}^{k} m_t^2\xi_t/\lambda_t \, (< \infty),$$

then for $N \to \infty$, $(\sqrt{N/v})(U - \theta)$ has a limiting $N(0, 1)$ distribution, where

$$\begin{aligned} \xi_t &= E[h(X_{1i_{11}}, \ldots, X_{1i_{1m_1}}; \ldots; X_{ki_{k1}}, \ldots, X_{ki_{km_k}}) \\ &\quad \times h(X_{1j_{11}}, \ldots, X_{1j_{1m_1}}; \ldots; X_{kj_{k1}}, \ldots, X_{kj_{km_k}})] - \theta^2, \end{aligned}$$

and the two sets $\{i_{s1}, \ldots, i_{sm_s}\}$, $\{j_{s1}, \ldots, j_{sm_s}\}$ have only one integer in common if $s = t$, and for all $s \neq t$, have no integers in common.

[Cf., Randles and Wolfe (1979), Hoeffding (1948).]

§2.31. Wald–Wolfowitz Theorem

Let $(X_1, \ldots, X_n)'$ be a random vector whose possible values are the $n!$ permutations of n real numbers $(c_{n1}, \ldots, c_{nn})$ which occur with the same probability $1/n!$. Define the linear statistic $L_n = \sum_{i=1}^n a_{ni} X_i$, where $a_{n1}, \ldots, a_{nn}$ are certain real numbers. Without loss of generality suppose that $\sum_{i=1}^n a_{ni} = 0$, $\sum_{i=1}^n c_{ni} = 0$. Suppose

$$\lim_{n\to\infty} \left[\sum_{i=1}^n (a_{ni})^r \Big/ n\right] \Big/ \left[\sum_{i=1}^n (a_{ni})^2 \Big/ n\right]^{r/2} = O(1),$$

and

$$\lim_{n\to\infty} \left[\sum_{i=1}^n (c_{ni})^r \Big/ n\right] \Big/ \left[\sum_{i=1}^n (c_{ni})^2 \Big/ n\right]^{r/2} = O(1),$$

for all $r = 1, 2, \ldots$. Then $[L_n/\sigma(L_n)]$ has, for $n \to \infty$, a limiting $N(0, 1)$ distribution, where

$$\sigma^2(L_n) = \frac{\left[\sum_{i=1}^n (c_{ni})^2 \sum_{j=1}^n (a_{nj})^2\right]}{(n-1)}.$$

For example, for Spearman's test (see §1.44(A)(iv)), under the null hypothesis, we may write $L_n = S/\sigma_{H_0}(S)$, where

$$a_{ni} = \frac{6}{n\sqrt{n-1}}\left[\frac{2i}{n+1} - 1\right], \qquad X_i = \left[R_i - \frac{(n+1)}{2}\right],$$

$$(c_{n1}, \ldots, c_{nn}) = \left(-\frac{(n-1)}{2}, -\frac{(n-3)}{2}, \ldots, 0, \ldots, \frac{(n-3)}{2}, \frac{(n-1)}{2}\right),$$

and we have used the property $\sum_{i=1}^n X_i = 0$ (with probability one). Then $L_n/\sigma(L_n)$ has, for $n \to \infty$, a limiting $N(0, 1)$ distribution, where $\sigma^2(L_n) = 1$.

See also §1.44.

[Cf. Wilks (1962).]

§2.32. Chernoff–Savage's for R-Statistics

Let $X_1, \ldots, X_{n_1}$ and $Y_1, \ldots, Y_{n_2}$ be independent samples from continuous distributions F and G, respectively. Let R_i denote the rank of Y_i in the pooled sample $(X_1, \ldots, X_{n_1}, Y_1, \ldots, Y_{n_2})$. Define the linear rank statistic $S_{n_1 n_2} = \sum_{i=1}^{n_2} a_N(R_i)$, where $N = n_1 + n_2$, and we consider two classes of scores $a_N(i)$:

(A) $$a_N(i) = \rho(i/(N+1)), \qquad 1 \leqslant i \leqslant N,$$

where ρ is nonconstant.

(B) $$a_N(i) = E[\rho(U_{(i)})], \qquad 1 \leqslant i \leqslant N,$$

where $U_{(i)}$ is the ith order statistic in a random sample of size N from the uniform [0, 1] distribution, and $\rho(x)$ is nondecreasing and nonconstant.

Suppose for all N, $\lambda_0 \leqslant n_2/N \leqslant (1 - \lambda_0)$ for some $0 < \lambda_0 < \frac{1}{2}$, where λ_0 is independent of N. Set $n_1/N = \lambda_N$, and define $H(x) = \lambda_N F(x) + (1 - \lambda_N)G(x)$. (Chernoff–Savage (also Hájek and Šidák).) If for $\rho(x)$ in class (A) or (B), $|(\partial/\partial x)^i \rho(x)| \leqslant K[x(1 - x)]^{-i-1/2+\delta}$, $i = 0, 1, 2$, for some $\delta > 0$, then $(S_{n_1 n_2} - \mu_{n_1 n_2})/\sigma_{n_1 n_2}$, for $N \to \infty$, has a limiting $N(0, 1)$ distribution, where

$$\mu_{n_1 n_2} = n_2 \int_{-\infty}^{\infty} \rho(H(x))\, dG(x),$$

$$\sigma^2_{n_1 n_2} = 2n_1(1 - \lambda_N)\{\lambda_N C_1 + (1 - \lambda_N)C_2\},$$

$$C_1 = \iint\limits_{-\infty < x < y < \infty} G(x)[1 - G(y)]\rho'(H(x))\rho'(H(y))\, dF(x)\, dF(y),$$

$$C_2 = \iint\limits_{-\infty < x < y < \infty} F(x)[1 - F(y)]\rho'(H(x))\rho'(H(y))\, dG(x)\, dG(y),$$

provided $\sigma^2_{n_1 n_2} > 0$.

For example, for the Wilcoxon–Mann–Whitney test (see §1.44(B)(i)), we may take

$$S_{n_1 n_2} = W_0/(N + 1) = \sum_{i=1}^{n_2} R_i/(N + 1),$$

and set $a_N(i) = i/(N + 1)$, $\rho(i) = i$, $\rho(x) = x$, $\rho'(x) = 1$. Under the null hypothesis,

$$\mu_{n_1 n_2} = n_2/2, \qquad C_1 = C_2 = \iint\limits_{-\infty < x < y < \infty} F(x)[1 - F(y)]\, dF(x)\, dF(y) = \frac{1}{24},$$

$\sigma^2_{n_1 n_2} = n_1 n_2/(12N)$. Here we may choose $K = 1$, $\delta = \frac{1}{2}$.

See also §1.44.

[Cf., Hájek and Šidák (1967), Chernoff and Savage (1958).]

§2.33. Miller's for Jackknife Statistics

(i) Let $X_1, \ldots, X_n$ be independent identically distributed random variables each with mean μ and variance σ^2 assumed finite. Let f be a real-valued function defined on the real line which, in the neighborhood of μ, has a bounded second derivative and $f'(\mu) > 0$. Define

$$\bar{X}_i = \sum_{\substack{j=1 \\ j \neq i}}^{n} X_j/(n-1), \qquad \bar{X} = \sum_{i=1}^{n} X_i/n, \qquad f(\mu) \equiv \theta,$$

$$\tilde{\theta}_i = nf(\bar{X}) - (n-1)f(\bar{X}_i), \qquad i = 1, \ldots, n,$$

and the jackknife estimator of θ:

$$\tilde{\theta}_. = nf(\bar{X}) - \frac{(n-1)}{n} \sum_{i=1}^{n} f(\bar{X}_i).$$

Then $[\sqrt{n}(\tilde{\theta}_. - \theta)/\sigma f'(\mu)]$ has, for $n \to \infty$, a limiting $N(0, 1)$ distribution.

(ii) Same notation and conditions as (i) except suppose that f has only a continuous first derivative in the neighborhood of μ. Then $\sum_{i=1}^{n} (\tilde{\theta}_i - \tilde{\theta}_.)^2/(n-1)$ converges in probability to $\sigma^2(f'(\mu))^2$.

(iii) Let $X_1, \ldots, X_n$ be independent identically distributed random variables each with mean μ, variance σ^2 and kurtosis γ_2 assumed finite: $0 < 2 + \gamma_2 < \infty$.

Set $\theta = \ln \sigma^2$, and define

$$S^2 = \sum_{i=1}^{n} (X_i - \bar{X})^2/(n-1), \qquad \bar{X}_i = \sum_{\substack{j=1 \\ j \neq i}}^{n} X_j/(n-1),$$

$$S_{-i}^2 = \sum_{\substack{j=1 \\ j \neq i}}^{n} (X_j - \bar{X}_i)^2/(n-2),$$

$$\tilde{\theta}_i = n \ln S^2 - (n-1) \ln S_{-i}^2,$$

$$\tilde{\theta}_. = n \ln S^2 - \frac{(n-1)}{n} \sum_{i=1}^{n} \ln S_{-i}^2.$$

Then

$$\frac{\sqrt{n}(\tilde{\theta}_. - \theta)}{\sqrt{\sum_{i=1}^{n} (\tilde{\theta}_i - \tilde{\theta}_.)^2/(n-1)}}$$

has, for $n \to \infty$ a limiting $N(0, 1)$ distribution. We note that we have used the unbiased estimator S^2 of σ^2, rather than using the biased estimator, by dividing by $(n-1)$ rather than by n.

(iv) Same notation as in (iii), $\sum_{i=1}^{n} (\tilde{\theta}_i - \tilde{\theta}_.)^2/(n-1)$ converges in probability to $(2 + \gamma_2)$.

See also §1.41.

[Cf., Miller (1964, 1968, 1974). See also Manoukian (1986).]

PART 2

STATISTICAL DISTRIBUTIONS

CHAPTER 3

Distributions

§3.1. Binomial

Consider n independent trials such that, in each trial, there is a probability p that an outcome E occurs. Then the probability that the random variable X, representing the number of trials in which E occurs, is equal to x is

$$P[X = x] = \binom{n}{x}(p)^x(1-p)^{n-x} \equiv b(x; n, p), \qquad x = 0, 1, \ldots, n,$$

$$\binom{n}{x} \equiv \frac{n!}{x!(n-x)!}, \qquad n! = n(n-1)\ldots 1, \; 0! = 1,$$

$$\Phi(t) = (1 - p + pe^{it})^n,$$

$$E[X] = np, \qquad \sigma^2(X) = np(1-p),$$

$$\gamma_1 = \frac{(1-2p)}{\sqrt{np(1-p)}}, \qquad \gamma_2 = \frac{1}{np(1-p)} - \frac{6}{n},$$

$$P[X \leqslant x] = \sum_{k=0}^{x} b(k; n, p) \equiv B(x; n, p),$$

$$B(x; n, p) = 1 - B(n - x - 1; n, 1 - p).$$

§3.2. Multinomial

Consider n independent trials such that in each trial one of k mutually independent events $E_1, \ldots, E_k$ occurs, with p_i denoting the probability that the event E_i occurs at each trial, and $\sum_{i=1}^{k} p_i = 1$. Then the joint probability mass function of the random variables $X_1, \dots, X_k$, representing the number of times the events $E_1, \dots, E_k$, respectively, occur in the n trials, is

$$P[X_1 = x_1, \dots, X_k = x_k] = \frac{n!}{x_1! \dots x_k!}(p_1)^{x_1} \dots (p_k)^{x_k},$$

$$x_i = 0, 1, \dots, n, \qquad i = 1, \dots, k, \qquad \sum_{i=1}^{k} x_i = n,$$

$$E[X_i] = np_i, \qquad \sigma^2(X_i) = np_i(1 - p_i), \qquad i = 1, \dots, k,$$

$$\mathrm{Cov}[X_i, X_j] = -np_ip_j, \qquad i \neq j.$$

§3.3. Geometric

Consider a series of independent trials, with p denoting the probability that an outcome E occurs in each trial. Then the probability that the random variable X, representing the number of consecutive trials needed so that the event E occurs for the first time at the last trial, is equal to x is given by

$$P[X = x] = (1 - p)^{x-1}p, \qquad x = 1, 2, \dots,$$

$$\Phi(t) = p[e^{-it} - (1 - p)]^{-1},$$

$$E[X] = \frac{1}{p}, \qquad \sigma^2(X) = \frac{1 - p}{p^2},$$

$$\gamma_1 = \frac{2 - p}{\sqrt{1 - p}}, \qquad \gamma_2 = 6 + \frac{p^2}{1 - p}.$$

§3.4. Pascal Negative Binomial

Consider a series of independent trials, with p denoting the probability that an outcome E occurs in each trial. Then the probability that the random variable X, representing the number of consecutive trials needed so that the event E occurs for the sth time at the last trial, is equal to x is given by

$$P[X = x] = \binom{x - 1}{s - 1}(p)^s(1 - p)^{x-s}, \qquad x = s, s + 1, \dots,$$

$$\Phi(t) = [p(e^{-it} - (1 - p))^{-1}]^s,$$

$$E[X] = \frac{s}{p}, \qquad \sigma^2(X) = \frac{s(1 - p)}{p^2},$$

$$\gamma_1 = \frac{2 - p}{\sqrt{s(1 - p)}}, \qquad \gamma_2 = \frac{6}{s} + \frac{p^2}{s(1 - p)}.$$

§3.5. Hypergeometric

Suppose a lot contains N items, of which D are defective. If a sample of n items is chosen, without replacement, from the lot, then the probability mass function of the random variable X, representing the number of defective items in the sample, is

$$P[X = x] = \binom{D}{x}\binom{N-D}{n-x}\bigg/\binom{N}{n},$$

$$x = a, a+1, \ldots, b; \quad a = \max[0, D+n-N], \quad b = \min[D, n],$$

$$E[X] = \frac{nD}{N}, \qquad \sigma^2(X) = \frac{nD}{N}\left(1 - \frac{D}{N}\right)\left(\frac{N-n}{N-1}\right).$$

§3.6. Poisson

The probability mass function of the Poisson distribution of parameter λ is defined by

$$P[X = x] = \frac{(\lambda)^x}{x!}e^{-\lambda}; \qquad \lambda > 0, \quad x = 0, 1, 2, \ldots,$$

$$\Phi(t) = \exp[\lambda(e^{it} - 1)],$$

$$E[X] = \lambda, \qquad \sigma^2(X) = \lambda,$$

$$\gamma_1 = \lambda^{-1/2}, \qquad \gamma_2 = \lambda^{-1},$$

$$P[X \leqslant x] = \sum_{k=0}^{x} \frac{(\lambda)^k}{k!}e^{-\lambda} \equiv F(x; \lambda).$$

§3.7. Wilcoxon's Null (One-Sample)

A random varible W is said to be distributed according to Wilcoxon's null one-sample distribution of sample size n if

$$P[W = x] = N(n, x)/2^n, \qquad x = 0, 1, \ldots, n(n+1)/2,$$

$N(n, 0) = 1$, and for $x \neq 0$, $N(n, x)$ denotes the number of subsets $\subseteq \{1, \ldots, n\}$ of distinct natural numbers such that the sum of the natural numbers in each subset is equal to x. [For example, for $n = 4$, $x = 5$, the subsets $\subseteq \{1, 2, 3, 4\}$, such that the sum of the natural numbers in each subset is equal to 5, are $\{1, 4\}$, $\{2, 3\}$. Hence $N(4, 5) = 2$.]

$$E[W] = \frac{n(n+1)}{4}, \qquad \sigma^2(W) = \frac{n(n+1)(2n+1)}{24}.$$

See also §§1.44(A)(ii) and 2.29(i).

§3.8. Wilcoxon–(Mann–Whitney)'s Null (Two-Sample)

A random variable W_0 is said to be distributed according to Wilcoxon's null two-sample distribution with sample sizes n_1 and n_2, respectively, if

$$P[W_0 = x] = t(x; n_2, N)\Big/\binom{N}{n_2},$$

$$N = n_1 + n_2,$$

$$x = \frac{n_2(n_2+1)}{2}, \frac{n_2(n_2+1)}{2} + 1, \ldots, \frac{n_2(N+n_1+1)}{2},$$

and $t(x; n_2, N)$ denotes the number of all subsets $\{t_1, \ldots, t_{n_2}\} \subset \{1, \ldots, N\}$, such that $t_1 < t_2 < \cdots < t_{n_2}$ and $t_1 + t_2 + \cdots + t_{n_2} = x$. [For example, for $n_1 = n_2 = 3$, $x = 12$, the subsets are $\{1, 5, 6\}$, $\{2, 4, 6\}$, $\{3, 4, 5\}$, and hence $t(12; 3, 6) = 3$.] The random variable W_0 is distributed symmetrically about the point $n_2(N+1)/2$.

$$E[W_0] = \frac{n_2(N+1)}{2}, \qquad \sigma^2(W_0) = \frac{n_1 n_2(N+1)}{12}.$$

The Mann–Whitney statistic is defined by $[W_0 - n_2(n_2+1)/2]$.
See also §§1.44(B)(i) and 2.29(ii).

§3.9. Runs

Consider a process giving rise to a sequence of two types of observations A and B. Let n_1 and n_2 denote the number of observations of types A and B, respectively, which are assumed to be fixed. We keep the data in the order they were produced. Let r denote the number of runs, that is, the number of groups where within each group all the observations are of the same type, and if the observations within a group are, say, of type A then the observations within the preceding group (if there is one) and within the proceeding one (if there is one) are of type B, etc. For example, if $(A, \ldots, A; B, \ldots, B; A, \ldots, A; B, \ldots, B)$ denotes the order that the observations are obtained, then the number of runs $r = 4$. Let R represent the random variable for the number of runs. Under the hypothesis that the process which generates the sequence of observations is a random process, we have with $n = n_1 + n_2$:

$$P[R = r] = 2\binom{n_1 - 1}{\frac{r}{2} - 1}\binom{n_2 - 1}{\frac{r}{2} - 1}\Big/\binom{n}{n_1}, \qquad r = \text{even},$$

$$P[R = r] = \left[\binom{n_1 - 1}{\frac{r-1}{2}}\binom{n_2 - 1}{\frac{r-3}{2}} + \binom{n_1 - 1}{\frac{r-3}{2}}\binom{n_2 - 1}{\frac{r-1}{2}}\right]\Big/\binom{n}{n_1},$$

$$r = \text{odd},$$

$$E[R] = 1 + \frac{2n_1 n_2}{n}, \qquad \sigma^2(R) = \frac{2n_1 n_2(2n_1 n_2 - n)}{n^2(n-1)}.$$

[Cf., Manoukian (1986).]

§3.10 Pitman–Fisher Randomization (One-Sample)

Let $z_1, \ldots, z_n$ be some fixed (nonzero) numbers not necessarily distinct. Define the collection of all 2^n n-vectors:

$$\Omega = \{(+|z_1|, +|z_2|, \ldots, +|z_n|)',$$
$$(-|z_1|, +|z_2|, \ldots, +|z_n|)', \ldots, (-|z_1|, \ldots, -(z_n))'\}$$

by attaching + or − signs to $|z_1|, \ldots, |z_n|$ in all possible ways. Let $\mathbf{S} = (S_1, \ldots, S_n)'$ be a random vector taking values from Ω with equal probabilities. Define the statistic $S = \sum_{i=1}^n S_i$. Let $\sum_{i=1}^n z_i = c$, and let q denote the number of S-values $\geqslant c$. Then

$$P[S \geqslant c] = q/2^n.$$

[For example, if $z_1 = 2, z_2 = 3, z_3 = 3, z_4 = -5$. Then $c = 3$, and $q = 7$. That is, $P[S \geqslant 3] = \frac{7}{16}$.]

$$E[S^{2r+1}] = 0, \qquad r = 0, 1, \ldots,$$

$$E[S^2] = \sum_{i=1}^n z_i^2,$$

$$E[S^4] = \left[3 - \frac{2}{n}\frac{\sum_{i=1}^n z_i^4/n}{\left(\sum_{i=1}^n z_i^2/n\right)^2}\right]\left(\sum_{i=1}^n z_i^2\right)^2.$$

See also §1.43 (and §1.3).
[Cf., Manoukian (1986).]

§3.11. Pitman's Permutation Test of the Correlation Coefficient

Let $(x_1, y_1), \ldots, (x_n, y_n)$ denote n pairs of fixed numbers, and without loss of generality suppose $\sum_{i=1}^n x_i = 0, \sum_{i=1}^n y_i = 0$. We fix the order of $x_1, \ldots, x_n$, and we consider all $n!$ pairings of the y's with the x's to generate the collection of all $n!$ n-tuples:

$$\Omega = \{((x_1, y_1), \ldots, (x_n, y_n))',$$
$$((x_1, y_2), (x_2, y_1), \ldots, (x_n, y_n))', \ldots, ((x_1, y_n), \ldots, (x_n, y_1))'\}.$$

Let $((x_1, W_1), \ldots, (x_n, W_n))'$ be a random variable taking any of the $n!$-values from Ω with equal probabilities. Define the statistic:

$$S = \sum_{i=1}^{n} x_i W_i.$$

Let $\sum_{i=1}^{n} x_i y_i = c$, and let q denote the number of S-values $\geqslant c$. Then

$$P[S \geqslant c] = q/n!.$$

[For example, for $n = 4$, $(x_1, y_1) = (-2, -1)$, $(x_2, y_2) = (-1, -1)$, $(x_3, y_3) = (0, -1)$, $(x_4, y_4) = (3, 3)$, $c = 12$ and $q = 6$. That is, $P[S \geqslant 12] = \frac{1}{4}$.] Since

$$\sum_{i=1}^{n} x_i^2. \qquad \sum_{i=1}^{n} W_i^2 = \sum_{i=1}^{n} y_i^2$$

are fixed numbers, we may also equivalently to S, consider the statistic:

$$S' = \frac{\left(\sum_{i=1}^{n} x_i W_i\right)}{\sqrt{\sum_{i=1}^{n} x_i^2 \sum_{j=1}^{n} W_j^2}}.$$

Then

$$E[\sqrt{n}\, S'] = 0, \qquad E[nS'^2] = n/(n-1),$$

and for n large:

$$E[n^{3/2} S'^3] = n^{-1/2} \frac{\left(\sum_{i=1}^{n} x_i^3/n\right)\left(\sum_{i=1}^{n} y_i^3/n\right)}{\left(\sum_{i=1}^{n} x_i^2/n\right)^{3/2}\left(\sum_{i=1}^{n} y_i^2/n\right)^{3/2}},$$

$$E[n^2 S'^4] = 3 + \frac{1}{n}\left[3 - \frac{\left(\sum_{i=1}^{n} x_i^4/n\right)}{\left(\sum_{i=1}^{n} x_i^2/n\right)^2}\right]\left[3 - \frac{\left(\sum_{i=1}^{n} y_i^4/n\right)}{\left(\sum_{i=1}^{n} y_i^2/n\right)^2}\right].$$

See also §1.43 (and §1.3).
[Cf., Manoukian (1986).]

§3.12. Pitman's Randomization (Two-Sample)

Let $x_1, \ldots, x_{n_1}, x_{n_1+1}, \ldots, x_{n_1+n_2}$ be $N = n_1 + n_2$ fixed numbers, and generate the collection $\Omega = \{(z_1, \ldots, z_{n_2})\}$ of all n_2-tuples, $n_2 < N$, which may be selected from $(x_1, \ldots, x_N)$ such that all orderings of the elements in $(z_1, \ldots, z_{n_2})$ are considered equivalent. Clearly Ω contains $\binom{N}{n_2}^2$ n_2-tuples. Let $(Z_1, \ldots, Z_{n_2})$ denote a random variable taking values from Ω with equal probabilities. Define the statistic

$$S = \sum_{i=1}^{n_2} Z_i.$$

Let $\sum_{i=n_1+1}^{N} x_i = c$, and let q denote the number of S-values $\geqslant c$. Then

$$P[S \geqslant c] = q \bigg/ \binom{N}{n_2}.$$

For each n_2-tuple $(Z_1, \ldots, Z_{n_2})$, there exists uniquely a $N - n_2 \equiv n_1$-tuple $(W_1, \ldots, W_{n_1})$, up to an ordering of its elements, such that the totality of the Z_i's and W_i's coincide with the fixed numbers $x_1, \ldots, x_N$. That is, in particular,

$$\sum_{i=1}^{n_2} Z_i + \sum_{i=1}^{n_1} W_i = \sum_{i=1}^{N} x_i, \qquad \sum_{i=1}^{n_2} Z_i^2 + \sum_{i=1}^{n_1} W_i^2 = \sum_{i=1}^{N} x_i^2$$

are fixed numbers. Hence, equivalently to S one may consider the statistic S':

$$S' = \frac{\bar{Z} - \bar{W}}{\sqrt{\dfrac{1}{n_1} + \dfrac{1}{n_2}}} \left(\frac{\sum_{i=1}^{n_2} (Z_i - \bar{Z})^2 + \sum_{i=1}^{n_1} (W_i - \bar{W})^2}{n_1 + n_2 - 2} \right)^{-1/2},$$

where $\bar{Z} = \sum_{i=1}^{n_2} Z_i/n_2$, $\bar{W} = \sum_{i=1}^{n_1} W_i/n_1$.

Let $\bar{S} = \sum_{i=1}^{n_2} Z_i/n_1$, $(n_1/n_2) \equiv \lambda$. Then

$$E[\bar{S}] = \sum_{i=1}^{N} x_i/N = \bar{x},$$

$$E[(\bar{S} - \bar{x})^2] = \frac{\lambda}{N}\left[\sum_{i=1}^{N} (x_i - \bar{x})^2/(N-1)\right],$$

$$E[(\bar{S} - \bar{x})^3] = \frac{\lambda(\lambda - 1)}{N^2}\left[N \sum_{i=1}^{N} (x_i - \bar{x})^3/(N-1)(N-2)\right],$$

$$E[(\bar{S} - \bar{x})^4] = \frac{\lambda}{(N-1)(N-2)(N-3)} \left\{ \left[(\lambda+1)^2 - 6(\lambda+1) + \frac{(\lambda+1)}{n_2} + 6 \right] \times \sum_{i=1}^{N} \frac{(x_i - \bar{x})^4}{N} + \left[3N\left(1 - \frac{1}{n_2}\right)\left(\lambda - \frac{1}{n_2}\right) \right] \left(\sum_{i=1}^{N} \frac{(x_i - \bar{x})^2}{N} \right)^2 \right\}.$$

See also §1.43 (and §1.3).
[Cf., Manoukian (1986).]

§3.13. Pitman's Randomization (k-Sample)

Consider a set of fixed numbers $x_{11}, \ldots, x_{1n_1}, x_{21}, \ldots, x_{2n_2}, \ldots, x_{k1}, \ldots, x_{kn_k}$, $\sum_{i=1}^{k} n_i = n$. Let Ω denote the collection of all partitions of the set of numbers $x_{11}, \ldots, x_{kn_k}$ into k groups of sizes $n_1, \ldots, n_k$, respectively. The elements in Ω will be denoted by $(\mathbf{z}_1, \ldots, \mathbf{z}_k)$, where $\mathbf{z}_i$ is n_i-dimensional, and all orderings of the components of a $\mathbf{z}_i$ are considered as equivalent. Clearly Ω contains $n!/n_1! \ldots n_k!$ elements. Let $(\mathbf{Z}_1, \ldots, \mathbf{Z}_k)$ denote a random variable taking values from Ω with equal probabilities. We introduce the statistic

$$S = \sum_{i=1}^{k} n_i \bar{Z}_{i.}^2$$

where $\bar{Z}_{i.} = \sum_{j=1}^{n_i} Z_{ij}/n_i$, and $\mathbf{Z}_i = (Z_{i1}, \ldots, Z_{in_i})'$. Let $\sum_{i=1}^{k} n_i \bar{x}_{i.}^2 = c$, and let q denote the number of S-values $\geqslant c$. Then

$$P[S \geqslant c] = q n_1! \ldots n_k!/n!.$$

Equivalently to the statistic S, we may consider the statistic

$$F' = \frac{\sum_{i=1}^{k} n_i(\bar{Z}_{i.} - \bar{Z}_{..})^2/(k-1)}{\sum_{i=1}^{k} \sum_{j=1}^{n_i} (Z_{ij} - \bar{Z}_{..})^2/(n-1)}.$$

Set $(x_{11}, \ldots, x_{kn_k}) \equiv (a_1, \ldots, a_n)$, $\bar{a} = \sum_{i=1}^{n} a_i/n$. Then

$$E[F'] = 1,$$

and for large $n_1, \ldots, n_k$

$$E[(F'-1)^2] = \frac{2}{k-1}\left\{1 + \frac{1}{n}\left[3 - \frac{\sum_{i=1}^{n} (a_i - \bar{a})^4/n}{\left(\sum_{i=1}^{n} (a_i - \bar{a})^2/n\right)^2}\right]\right\}.$$

See also §1.43 (and §1.3).
[Cf., Manoukian (1986).]

§3.14. Kolmogorov–Smirnov's Null (One-Sample)

Let $X_1, \ldots, X_n$ be independent identically distributed random variables each with a continuous distribution $F(x)$. Let $\hat{F}_n(x)$ denote the sampling distribution, that is, $\hat{F}_n(x) = [\text{number of } X_i \leqslant x]/n$. Define

$$D_n = \sup_x |\hat{F}_n(x) - F(x)|,$$

$$D_n^+ = \sup_x [\hat{F}_n(x) - F(x)],$$

$$D_n^- = \sup_x [F(x) - \hat{F}_n(x)].$$

Then

$$P\left[D_n < \frac{k}{n}\right] = \frac{n!}{n^n} U(k, n), \qquad k = 1, \ldots, n-1,$$

where $U(i, j+1)$, $i = 1, \ldots, 2k-1$, $j = 0, 1, \ldots, n-1$, satisfy

$$U(i, j+1) = \sum_{t=1}^{i+1} \frac{U(t, j)}{(i+1-t)!},$$

subject to

$$U(t, 0) = 0, \quad t = 1, \ldots, k-1; \qquad U(k, 0) = 1; \qquad U(t, j) = 0, \quad t \geqslant j + k.$$

Also,

$$P[D_n^+ \leqslant d] = P[D_n^- \leqslant d]$$

$$= 1 - d \sum_{j=0}^{[n-nd]} \binom{n}{j} \left(\frac{j}{n} + d\right)^{j-1} \left(1 - \frac{j}{n} - d\right)^{n-j}, \quad 0 < d < 1,$$

where $[x]$ denotes the largest integer $\leqslant x$.

See also §§1.44 and 2.22.

[Cf., Wilks (1962), Durbin (1973), Manoukian (1986).]

§3.15. Kolmogorov–Smirnov's Null (Two-Sample)

Let $X_1, \ldots, X_{n_1}, Y_1, \ldots, Y_{n_2}$ be independent identically distributed random variables each with a continuous distribution $F(x)$. Define the sampling distributions

$$\hat{F}_{n_1}(x) = [\text{number of } X_i \leqslant x]/n_1, \qquad \hat{G}_{n_2}(x) = [\text{number of } Y_i \leqslant x]/n_2.$$

We consider only the case where $n_1 = n_2 \equiv n$. Define

$$D_{n,n} = \sup_x |\hat{F}_n(x) - \hat{G}_n(x)|,$$

$$D_{n,n}^+ = \sup_x [\hat{F}_n(x) - \hat{G}_n(x)],$$

$$D_{n,n}^- = \sup_x [\hat{G}_n(x) - \hat{F}_n(x)].$$

Let h be a positive integer $0 < h < n$. Then

$$P\left[D_{n,n} \geqslant \frac{h}{n}\right] = 2 \sum_{j=0}^{[n/h]-1} (-1)^j \binom{2n}{n-(j+1)h} \Big/ \binom{2n}{n},$$

where $[x]$ is the largest integer $\leqslant x$. Also,

$$P\left[D_{n,n}^+ \geqslant \frac{h}{n}\right] = P\left[D_{n,n}^- \geqslant \frac{h}{n}\right] = \binom{2n}{n-h} \Big/ \binom{2n}{n}.$$

See also §§1.44 and 2.22.

[Cf., Wilks (1962), Durbin (1973), Manoukian (1986).]

§3.16. Uniform (Rectangular)

The probability density function is defined by

$$f(x) = \begin{cases} 0, & x < a, \\ (b-a)^{-1}, & a \leqslant x \leqslant b, \\ 0, & x > b, \end{cases}$$

$$\Phi(t) = \frac{[e^{itb} - e^{ita}]}{it(b-a)},$$

$$E[X] = \frac{a+b}{2}, \qquad \sigma^2(X) = \frac{(b-a)^2}{12},$$

$$\gamma_1 = 0, \qquad \gamma_2 = -\frac{6}{5}, \qquad \text{median} = \frac{a+b}{2}.$$

§3.17. Triangular

The probability density function is defined by

$$f(x) = \begin{cases} 0, & x \leqslant a, \\ 4(x-a)/(b-a)^2, & a < x \leqslant (a+b)/2, \\ 4(b-x)/(b-a)^2, & (a+b)/2 < x < b, \\ 0, & x \geqslant b, \end{cases}$$

$$\Phi(t) = 4[e^{ita/2} - e^{itb/2}]^2/t^2(b-a)^2,$$

$$E[X] = (a+b)/2, \qquad \sigma^2(X) = (b-a)^2/24,$$

$$\gamma_1 = 0, \qquad \gamma_2 = -3/5, \qquad \text{median} = (a+b)/2.$$

§3.18. Pareto

The probability density function is defined by

$$f(x) = ac^a/x^{a+1}, \qquad a > 0, \qquad c > 0, \qquad x \geqslant c,$$

$$E[X^r] = ac^r/(a-r), \qquad a > r,$$

$$\text{mode} = c, \qquad \text{median} = c2^{1/a}.$$

§3.19. Exponential

The probability density function is defined by

$$f(x) = \alpha e^{-\alpha x}, \qquad x \geqslant 0, \quad \alpha > 0,$$

$$\Phi(t) = \alpha/(\alpha - it),$$

$$E[X] = \alpha^{-1}, \qquad \sigma^2(X) = \alpha^{-2},$$

$$\gamma_1 = 2, \qquad \gamma_2 = 6,$$

$$\text{mode} = 0, \qquad \text{median} = \alpha^{-1} \ln 2.$$

We note that the exponential distribution has no "memory," that is,

$$P[X > x + x' | X > x'] = P[X > x].$$

§3.20. Erlang and Gamma

The probability density of the gamma distribution is defined by

$$f(x) = \frac{(\alpha)^k (x)^{k-1}}{\Gamma(k)} e^{-\alpha x}, \qquad \alpha > 0, \quad k > 0, \quad x \geqslant 0,$$

where $\Gamma(k)$ is the gamma function.

$$\Phi(t) = [\alpha/(\alpha - it)]^k,$$

$$E[X] = k/\alpha, \qquad \sigma^2(X) = k/\alpha^2,$$

$$\gamma_1 = 2k^{-1/2}, \qquad \gamma_2 = 6k^{-1},$$

$$\text{mode} = (k-1)/\alpha, \qquad k \geqslant 1.$$

If k is a natural number, then the gamma distribution is called the Erlang distribution.

§3.21. Weibull and Rayleigh

The probability density of the Weibull distribution is defined by

$$f(x) = \frac{\beta}{\alpha}\left(\frac{x-\gamma}{\alpha}\right)^{\beta-1} \exp\left[-\left(\frac{x-\gamma}{\alpha}\right)^{\beta}\right],$$

$$\gamma \leqslant x < \infty, \qquad \alpha > 0, \qquad \beta > 0, \qquad -\infty < \gamma < \infty,$$

$$E[X] = \gamma + \alpha\Gamma\left(\frac{\beta+1}{\beta}\right),$$

$$\sigma^2(X) = \alpha^2\left[\Gamma\left(\frac{\beta+2}{\beta}\right) - \Gamma^2\left(\frac{\beta+1}{\beta}\right)\right],$$

where $\Gamma(\beta)$ is the gamma function. For $\gamma = 0$, $\beta = 2$, the distribution is called that of Rayleigh.

§3.22. Beta

The probability density function is defined by

$$f(x) = \frac{\Gamma(\alpha+\beta)}{\Gamma(\alpha)\Gamma(\beta)}(x)^{\alpha-1}(1-x)^{\beta-1}, \qquad 0 \leqslant x \leqslant 1, \quad \alpha > 0, \quad \beta > 0,$$

$\Gamma(\alpha)$ is the gamma function.

$$E[X] = \frac{\alpha}{(\alpha+\beta)}, \qquad \sigma^2(X) = \frac{\alpha\beta}{(\alpha+\beta)^2(\alpha+\beta+1)}.$$

§3.23. Half-Normal

The probability density function is defined by

$$f(x) = \frac{2\theta}{\pi}\exp\left[-\left(\frac{\theta^2 x^2}{\pi}\right)\right], \qquad 0 \leqslant x < \infty,$$

$$E[X] = \frac{1}{\theta}, \qquad \sigma^2(X) = \left(\frac{\pi-2}{2}\right)\frac{1}{\theta^2},$$

$$\gamma_1 = \frac{(4-\pi)}{\theta^3}, \qquad \gamma_2 = \frac{3\pi^2-4\pi-12}{4\theta^4} - 3.$$

§3.24. Normal (Gauss)

The probability density is defined by

$$f(x) = \frac{1}{\sqrt{2\pi}\,\sigma}\exp\left[\frac{-(x-\mu)^2}{2\sigma^2}\right],$$

$$-\infty < x < \infty, \qquad -\infty < \mu < \infty, \qquad 0 < \sigma < \infty,$$

$$\Phi(t) = \exp[it\mu - \tfrac{1}{2}\sigma^2 t^2],$$

$$E[X] = \mu, \qquad \sigma^2(X) = \sigma^2,$$

$$\gamma_1 = 0, \qquad \gamma_2 = 0,$$

$$\text{mode} = \mu, \qquad \text{median} = \mu,$$

$$P[X \leqslant x] = P\left[Z \leqslant \frac{x-\mu}{\sigma}\right] \equiv \phi\left(\frac{x-\mu}{\sigma}\right),$$

$$\phi(a) = \frac{1}{\sqrt{2\pi}}\int_{-\infty}^{a} e^{-x^2/2}\,dx, \qquad Z = (X-\mu)/\sigma.$$

Z is called a standard normal variable ($E[Z] = 0$, $\sigma^2(Z) = 1$), and $\phi(a)$ is called the standard normal distribution. The $(1 - \alpha)$th quantile of the standard normal distribution will be denoted by Z_α, that is, $\phi(Z_\alpha) = 1 - \alpha$.

§3.25. Cauchy

The probability density function is defined by

$$f(x) = b(\pi[(x - a)^2 + b^2])^{-1},$$

$$-\infty < x < \infty, \qquad -\infty < a < \infty, \qquad 0 < b < \infty,$$

$$\text{mode} = a, \qquad \text{median} = a,$$

The moments $E[X^r]$, $r = 1, 2, \ldots$, do not exist.

§3.26. Lognormal

The probability density function is defined by

$$f(x) = \frac{1}{x\sigma\sqrt{2\pi}} \exp\left[\frac{-(\ln x - \mu)^2}{2\sigma^2}\right],$$

$$0 < x < \infty, \qquad -\infty < \mu < \infty, \qquad 0 < \sigma < \infty,$$

$$E[X] = \exp\left[\mu + \frac{\sigma^2}{2}\right], \qquad \sigma^2(X) = e^{(2\mu+\sigma^2)}[e^{\sigma^2} - 1],$$

$$\text{mode} = e^{(\mu-\sigma^2)}, \qquad \text{median} = e^{\mu}.$$

§3.27. Logistic

The probability density function is defined by

$$f(x) = \frac{\exp[(x - \theta)/\beta]}{\beta(1 + \exp[(x - \theta)/\beta])^2},$$

$$-\infty < x < \infty, \quad -\infty < \theta < \infty, \quad 0 < \beta < \infty,$$

$$\Phi(t) = \frac{(2e^{it\theta}\pi\beta t)}{(e^{\pi\beta t} - e^{-\pi\beta t})},$$

$$E[X] = \theta, \qquad \sigma^2(X) = \beta^2\pi^2/3,$$

$$\text{mode} = \theta, \qquad \text{median} = \theta,$$

$$\gamma_1 = 0, \qquad \gamma_2 = 1.2.$$

§3.28. Double-Exponential

The probability density is defined by

$$f(x) = (2\beta)^{-1} \exp\left[\frac{-|x-\theta|}{\beta}\right],$$

$$-\infty < x < \infty, \quad -\infty < \theta < \infty, \quad 0 < \beta < \infty,$$

$$\Phi(t) = e^{it\theta}[1 + \beta^2 t^2]^{-1},$$

$$E[X] = \theta, \qquad \sigma^2(X) = 2\beta^2,$$

$$\text{mode} = \theta, \qquad \text{median} = \theta,$$

$$\gamma_1 = 0, \qquad \gamma_2 = 3.$$

§3.29. Hyperbolic-Secant

The probability density is defined by

$$f(x) = \frac{1}{\sigma}\left(\exp\left[\frac{\pi(x-\mu)}{2\sigma}\right] + \exp\left[\frac{-\pi(x-\mu)}{2\sigma}\right]\right)^{-1},$$

$$-\infty < x < \infty, \quad -\infty < \mu < \infty, \quad 0 < \sigma < \infty,$$

$$\Phi(t) = \frac{2e^{it\mu}e^{-\sigma t}}{(1 + e^{-2\sigma t})},$$

$$E[X] = \mu, \qquad \sigma^2(X) = \sigma^2,$$

$$\gamma_1 = 0, \qquad \gamma_2 = 2,$$

$$\text{mode} = \mu, \qquad \text{median} = \mu.$$

§3.30. Slash

The probability density function is defined by

$$f(x) = \frac{\lambda}{\sqrt{2\pi}} \frac{1 - \exp\{-[(x-\theta)^2/2\lambda^2]\}}{(x-\theta)^2},$$

$$-\infty < x < \infty, \quad -\infty < \theta < \infty, \quad 0 < \lambda < \infty,$$

For

$$x \to \theta, \qquad f(x) \to 1/(2\lambda\sqrt{2\pi}),$$

$$\text{mode} = \theta, \qquad \text{median} = \theta.$$

The moments $E[X^r]$, $r = 1, 2, \ldots$, do not exist.
[Cf., Hoaglin *et al.* (1983).]

§3.31. Tukey's Lambda

It is defined in terms of the inverse distribution function:

$$x = F^{-1}(p) = \lambda_1 + \frac{p^{\lambda_3} - (1-p)^{\lambda_3}}{\lambda_2}, \qquad 0 < p < 1,$$

where the parameters $\lambda_1, \lambda_2, \lambda_3$ are such that $-\infty < \lambda_1 < \infty$, $\lambda_2 \neq 0$, $\lambda_3 \neq 0$, $\lambda_3/\lambda_2 > 0$. For $\lambda_2 < 0$, $\lambda_3 < 0$: $-\infty < x < \infty$; for $\lambda_2 > 0$, $\lambda_3 > 0$: $\lambda_1 - \lambda_2^{-1} < x < \lambda_1 + \lambda_2^{-1}$. The probability density function may be defined by:

$$f(F^{-1}(p)) = [\lambda_3(p^{\lambda_3 - 1} + (1-p)^{\lambda_3 - 1})/\lambda_2]^{-1}, \quad \text{(Ramberg and Schmeiser)}$$

$$E[X] = \lambda_1,$$

$$\sigma^2(X) = 2[(2\lambda_3 + 1)^{-1} - \Gamma^2(\lambda_3 + 1)(\Gamma(2\lambda_3 + 2))^{-1}]\lambda_2^{-2},$$

$$\lambda_3 \neq -\tfrac{1}{2}, \quad \lambda_3 > -1,$$

$$\gamma_3 = 0,$$

$$\gamma_4 = \tfrac{1}{2}\left[(4\lambda_3 + 1)^{-1} - \frac{4\Gamma(\lambda_3 + 1)\Gamma(3\lambda_3 + 1)}{\Gamma(4\lambda_3 + 2)} + \frac{3\Gamma^2(2\lambda_3 + 1)}{\Gamma(4\lambda_3 + 2)}\right]$$

$$\times [(2\lambda_3 + 1)^{-1} - \Gamma^2(\lambda_3 + 1)(\Gamma(2\lambda_3 + 2))^{-1}]^{-2} - 3,$$

$$\lambda_3 \neq -\tfrac{1}{4}, \quad \lambda_3 > -\tfrac{1}{3}.$$

median $= \lambda_1$, where $\Gamma(z)$ is the gamma function. [An asymmetry may be introduced in the distribution by introducing in turn a new parameter $\lambda_4 \neq \lambda_3$, and upon writing $F^{-1}(p) = \lambda_1 + (p^{\lambda_3} - (1-p)^{\lambda_4})/\lambda_2$ [Ramberg and Schmeiser (1974)].]

[Cf., Ramberg and Schmeiser (1972, 1974).]

§3.32. Exponential Family

I. *One-dimensional Parameter*. Let X be a random variable with probability density or probability mass $f(x; \theta)$ depending on a parameter θ, where $\theta \in \Omega \subseteq R^1$. Then $f(x; \theta)$ is said to belong to the exponential family if it is of the form:

$$f(x; \theta) = a(\theta)\exp[b(\theta)t(x)]h(x),$$

where $a(\theta) > 0$ for $\theta \in \Omega$, and the set of positivity of $f(x; \theta)$ is independent of θ, that is the set $\{x : f(x; \theta) > 0\}$ is independent of θ. We note that if $X_1, \ldots, X_n$ are independent identically distributed random variables each with a probability density or probability mass $f(x; \theta)$ as given above, then the joint probability density or probability mass function of $X_1, \ldots, X_n$ is

$$\prod_{i=1}^{n} f(x_i; \theta) = (a(\theta))^n \exp\left[b(\theta)\sum_{i=1}^{n} t(x_i)\right]\prod_{i=1}^{n} h(x_i),$$

and is of the exponential family type.

II. *k-dimensional Parameter*. Let $X_1, \ldots, X_n$ be independent identically distributed random variables. Then the joint probability density or probability mass of $X_1, \ldots, X_n$ is said to belong to a k-parameter exponential family if it is of the form:

$$f(\mathbf{x}; \boldsymbol{\theta}) = a(\boldsymbol{\theta}) \exp\left[\sum_{i=1}^{k} c_i(\boldsymbol{\theta}) t_i(\mathbf{x})\right] h(\mathbf{x}),$$

where $a(\boldsymbol{\theta}) > 0$ for $\boldsymbol{\theta} = (\theta_1, \ldots, \theta_k)' \in \Omega \subseteq R^k$, $\mathbf{x} = (x_1, \ldots, x_n)'$ and the set of positivity of $f(\mathbf{x}; \boldsymbol{\theta})$ is independent of $\boldsymbol{\theta}$.

[Cf., Lehmann (1959).]

§3.33. Exponential Power

A continuous random variable X with probability density function:

$$f(x) = \frac{c(\delta)}{\sigma} \exp[-(Q(x; \mu, \sigma, \delta))^{1/\delta}], \qquad -\infty < x < \infty,$$

where

$$c(\delta) = \frac{1}{2}\sqrt{\frac{\Gamma(3\delta)}{\Gamma(\delta)}}\frac{1}{\Gamma(1+\delta)},$$

$$Q(x; \mu, \sigma, \delta) = \sqrt{\frac{\Gamma(3\delta)}{\Gamma(\delta)}}\frac{|x-\mu|}{\sigma}, \qquad -\infty < \mu < \infty, \quad 0 < \sigma < \infty, \quad 0 < \delta \leqslant 1,$$

and $\Gamma(\delta)$ is the gamma function,

$$E[X] = \mu, \qquad \sigma^2(X) = \sigma^2,$$

$$\gamma_1 = 0, \qquad \gamma_2 = \frac{\Gamma(5\delta)\Gamma(\delta)}{(\Gamma(3\delta))^2} - 3.$$

The following identities are useful for considering special cases for δ:

$$\Gamma(n\delta) = (2\pi)^{(1-n)/2} n^{n\delta-(1/2)} \prod_{j=0}^{n-1} \Gamma\left(\delta + \frac{j}{n}\right),$$

where n is a positive integer,

$$\Gamma(a)\Gamma(1-a) = \pi/\sin \pi a, \qquad 0 < a < 1,$$

$$\Gamma(\tfrac{1}{2}) = \sqrt{\pi}.$$

The parameter δ is an alternative measure of peakedness. For $\delta \to 0$, $\delta = \frac{1}{2}$, $\delta = 1$, we have, respectively, the uniform, the normal and the double-exponential distributions.

[Cf., Box and Tiao (1973), Gradshteyn and Ryzhik (1965).]

§3.34. Pearson Types

Consider a continuous random variable X with a probability density $f(x)$ such that the set of positivity of $f(x)$ is given by the range $a < x < b$, where a and b may $-\infty$ and $+\infty$, respectively. We will suppose that the first moment of X exists and without loss of generality we set $\mathrm{E}[X] = 0$. If the probability density $f(x)$ satisfies a differential equation of the form:

$$\left[(C_0 + C_1 x + C_2 x^2)\frac{d}{dx} + (C_1 + x)\right] f(x) = 0, \qquad a < x < b,$$

then we will say that $f(x)$ is of a Pearson-type density. In the sequel k denotes a normalization constant determined in such a way that $\int_a^b f(x)\,dx = 1$.

Type I. Specified by the existence of two real roots of $C_0 + C_1 x + C_2 x^2 = 0$, of opposite signs, say $x_1, x_2, x_1 < 0 < x_2$:

$$f(x) = k\left(1 - \frac{x}{x_1}\right)^{m_1}\left(1 - \frac{x}{x_2}\right)^{m_2}, \qquad x_1 < x < x_2,$$

$$m_1 = (C_1 + x_1)/(C_2(x_2 - x_1)),$$

$$m_2 = -(C_1 + x_2)/(C_2(x_2 - x_1)), \qquad m_1 > -1, \quad m_2 > -1.$$

Type II. In reference to Type I, if $x_1 = -x_2 \equiv -x_0 < 0$, then $C_1 = 0$, $m_1 = m_2 = -1/2C_2$:

$$f(x) = k\left(1 - \frac{x^2}{x_0^2}\right)^{-1/2C_2}, \qquad -x_0 < x < x_0, \quad 1/2C_2 < 1.$$

Type III. Specified by $C_2 = 0$, $C_1 \neq 0$:

$$f(x) = k(C_0 + C_1 x)^m \exp[-x/C_1], \qquad m = \frac{1}{C_1}\left(\frac{C_0}{C_1} - C_1\right),$$

if $C_1 > 0$, then $-C_0/C_1 < x < \infty$, and if $C_1 < 0$, then $-\infty < x < -C_0/C_1$; $m > -1$.

Type IV. If $C_0 + C_1 x + C_2 x^2 = 0$ has no real roots, that is, $4C_0 C_2 - C_1^2 > 0$:

$$f(x) = k(C_0 + C_1 x + C_2 x^2)^{-1/2C_2} \exp\left[-\gamma \tan^{-1}\left(\frac{2C_2 x + C_1}{b}\right)\right],$$

$$-\infty < x < \infty,$$

$$\gamma = C_1(2C_2 - 1)/C_2 b, \qquad b = \sqrt{4C_0 C_2 - C_1^2}; \qquad 1/C_2 > 2.$$

Type V. Specified by the condition that $C_0 + C_1x + C_2x^2 = 0$ has two identical roots not equal to $-C_1$, that is,

$$C_0 + C_1x + C_2x^2 = C_2(x - B)^2, \qquad B \neq -C_1:$$

$$f(x) = k(x - B)^{-1/C_2} \exp\left[\frac{(C_1 + B)}{C_2(x - B)}\right],$$

if $(C_1 + B)/C_2 < 0$, $B < x < \infty$, and

$$f(x) = k(B - x)^{-1/C_2} \exp\left[\frac{(C_1 + B)}{C_2(x - B)}\right],$$

if $(C_1 + B)/C_2 > 0$, $-\infty < x < B$; $1/C_2 > 2$.

Type VI. Specified by the condition that $C_0 + C_1x + C_2x^2 = 0$ has real roots, say x_1, x_2, of the same sign, and for $x_1 < x_2 < 0$:

$$f(x) = k(x - x_1)^{m_1}(x - x_2)^{m_2}, \qquad x_2 < x < \infty,$$

with $0 < m_2 + 1 < -m_1 - 1$, and for $0 < x_1 < x_2$:

$$f(x) = k(x_1 - x)^{m_1'}(x_2 - x)^{m_2'}, \qquad -\infty < x < x_1,$$

with $0 < m_1' + 1 < -m_2' - 1$.

Type VII. If $C_1 = 0$, $C_0 > 0$, $C_2 > 0$:

$$f(k) = k(C_0 + C_2x^2)^{-1/2C_2}, \qquad -\infty < x < \infty,$$

$1/C_2 > 2$. (This may be obtained from Type IV by setting $C_1 = 0$, etc.)

The first four moments of X for the above distributions exist if in addition we require that: $1/C_2 > 5$ for Types IV, V, and VII;

$$0 < m_2 + 1 < -m_1 - 4 \quad \text{or} \quad 0 < m_1' + 1 < -m_2' - 4,$$

respectively, for Type VI, and with no other changes for Types I, II, and III. With these restrictions one may verify that

$$(x)^n(C_0 + C_1x + C_2x^2)f(x) = 0$$

for $x \to a$ and for $x \to b$, for $n = 0, 1, 2, 3$. Without loss of generality suppose $E[X^2] = 1$, in addition to the condition $E[X] = 0$. Also set $E[X^3] = \mu_3$, $E[X^4] = \mu_4$. Multiplying the differential equation for $f(x)$ from the left by $(x)^n$, integrating by parts and using the boundary conditions

$$(x)^n(C_0 + C_1x + C_2x^2)f(x) = 0$$

for $x \to a$ and for $x \to b$, for $n = 0, 1, 2, 3$, we see that the parameters C_0, C_1 and C_2 may be expressed in terms of μ_3 and μ_4:

$$C_0 = \frac{4\mu_4 - 3\mu_3^2}{10\mu_4 - 12\mu_3^2 - 18},$$

$$C_1 = \frac{\mu_3(\mu_4 + 3)}{10\mu_4 - 12\mu_3^2 - 18},$$

$$C_2 = \frac{2\mu_4 - 3\mu_3^2 - 6}{10\mu_4 - 12\mu_3^2 - 18}.$$

[Cf., Cramér (1974), Johnson and Kotz (1969–1972).]

§3.35. Chi-Square χ^2

The probability density of the chi-square distribution of ν degrees of freedom is defined by

$$f_{\chi^2}(x) = \frac{1}{2^{\nu/2}\Gamma(\nu/2)} x^{(\nu-2)/2} e^{-x/2}, \qquad x > 0,$$

where $\Gamma(\nu)$ is the gamma function.

$$\Phi(t) = [1 - 2it]^{-\nu/2},$$

$$E[\chi^2] = \nu, \qquad \sigma^2(\chi^2) = 2\nu,$$

$$\gamma_1 = \sqrt{8/\nu}, \qquad \gamma_2 = 12/\nu.$$

We will denote by $\chi_\alpha^2(\nu)$ the $(1 - \alpha)$th quantile, that is,

$$P[\chi^2 \leqslant \chi_\alpha^2(\nu)] = 1 - \alpha.$$

§3.36. Student's T

The probability density function of the Student distribution of ν degrees of freedom is defined by

$$f_T(x) = \frac{1}{\sqrt{\pi\nu}} \frac{\Gamma\left(\frac{\nu+1}{2}\right)}{\Gamma\left(\frac{\nu}{2}\right)} \left[1 + \frac{x^2}{\nu}\right]^{-(\nu+1)/2}, \qquad -\infty < x < \infty,$$

$$E[T] = 0,$$

$$\sigma^2(T) = \nu/(\nu - 2), \qquad \nu > 2,$$

$$\gamma_1 = 0,$$

$$\gamma_2 = 6/(\nu - 4), \qquad \nu > 4,$$

$$\text{mode} = 0, \qquad \text{median} = 0.$$

We will denote by $t_\alpha(\nu)$ the $(1 - \alpha)$th quantile, that is,

$$P[T \leqslant t_\alpha(\nu)] = 1 - \alpha.$$

§3.37. Fisher's F

The probability density of Fisher's F-distribution of ν_1 and ν_2 degrees of freedom is defined by

$$f_F(x) = \frac{\Gamma\left(\dfrac{\nu_1+\nu_2}{2}\right)}{\Gamma\left(\dfrac{\nu_1}{2}\right)\Gamma\left(\dfrac{\nu_2}{2}\right)} (\nu_1)^{\nu_1/2}(\nu_2)^{\nu_2/2} \frac{x^{(\nu_1/2)-1}}{(\nu_2+\nu_1 x)^{(\nu_1+\nu_2)/2}}, \qquad x > 0,$$

$$E[F] = \nu_2/(\nu_2 - 2), \qquad \nu_2 > 2,$$

$$\sigma^2(F) = \frac{2\nu_2^2(\nu_1+\nu_2-2)}{\nu_1(\nu_2-2)^2(\nu_2-4)}, \qquad \nu_2 > 4.$$

We will denote by $F_\alpha(\nu_1, \nu_2)$ the $(1-\alpha)$th quantile, that is,

$$P[F \leqslant F_\alpha(\nu_1, \nu_2)] = 1 - \alpha.$$

§3.38. Noncentral Chi-Square

The probability density of the noncentral chi-square distribution of ν degrees of freedom and a noncentrality parameter $\delta > 0$ is defined by

$$f_{\chi^2}(x; \delta, \nu) = e^{-\delta/2} \sum_{k=0}^{\infty} \frac{(\delta/2)^k}{k!} f_{\chi^2}(x; \nu + 2k), \qquad x > 0,$$

where $f_{\chi^2}(x; \nu + 2k)$ is the probability density of a chi-square distribution of $(\nu + 2k)$ degrees of freedom,

$$f_{\chi^2}(x; 0, \nu) \equiv f_{\chi^2}(x; \nu),$$

$$\Phi(t) = [1 - 2it]^{-\nu/2} \exp\left[\frac{it\delta}{1-2it}\right],$$

$$E[\chi^2(\delta; \nu)] = \nu + \delta.$$

§3.39. Noncentral Student

The probability density of the noncentral Student distribution of ν degrees of freedom and a noncentrality parameter μ is defined by

$$f_T(x; \mu, \nu) = \frac{1}{\sqrt{\pi\nu}\, 2^{(\nu+1)/2}\Gamma(\nu/2)} \int_0^\infty y^{(\nu-1)/2} \times \exp\{-\tfrac{1}{2}[y + (x\sqrt{y/\nu} - \mu)^2]\}\, dy, \qquad -\infty < x < \infty.$$

§3.40. Noncentral Fisher's F

The probability density of the noncentral Fisher F-distribution of ν_1 and ν_2 degrees of freedom and a noncentrality parameter $\delta > 0$ is defined by

$$f_F(x; \delta, \nu_1, \nu_2) = e^{-\delta/2} \sum_{k=0}^{\infty} \frac{(\delta/2)^k}{k!} f_F(x; \nu_1 + 2k, \nu_2), \qquad x > 0,$$

where $f_F(x; \nu_1 + 2k, \nu_2)$ is the probability density of an F-distribution of $(\nu_1 + 2k)$ and ν_2 degrees of freedom, $f_F(x; 0, \nu_1, \nu_2) \equiv f_F(x; \nu_1, \nu_2)$.

§3.41. Order Statistics

Let $X_1, \ldots, X_n$ be independent identically distributed random variables each with a continuous distribution $F(x)$ and probability density $f(x)$. Suppose $f(x) \neq 0$ only for $a < x < b$ (a and b may be $-\infty$ and $+\infty$). Let $Y_1 \leqslant \cdots \leqslant Y_n$ be the order statistics of $X_1, \ldots, X_n$. Then the joint probability density of the order statistics $Y_1, \ldots, Y_n$ is given by:

$$f(y_1, \ldots, y_n) = n!\, f(y_1) \ldots f(y_n),$$

for $a < y_1 < y_2 < \cdots < y_n < b$, and is zero otherwise. The joint probability density of a pair of order statistics Y_i, Y_j, with $i < j$, is given by

$$f_{ij}(y_i, y_j) = n! \frac{[1 - F(y_j)]^{n-j}}{(n-j)!} \frac{[F(y_j) - F(y_i)]^{j-i-1}}{(j-i-1)!} \frac{[F(y_i)]^{i-1}}{(i-1)!} f(y_i) f(y_j),$$

for $a < y_i < y_j < b$, and is zero otherwise. The probability density of the ith order statistic is given by

$$f_i(y_i) = \frac{n!}{(i-1)!\,(n-i)!} [F(y_i)]^{i-1} [1 - F(y_i)]^{n-i} f(y_i),$$

for $a < y_i < b$, and is zero otherwise (see also §§3.42, 3.43, 3.44, and 3.45).

§3.42. Sample Range

Let $X_1, \ldots, X_n$ be independent identically distributed random variables each with a continuous distribution $F(x)$, and probability density $f(x)$. Suppose $f(x) \neq 0$ only for $a < x < b$ (a and b may be $-\infty$ and $+\infty$). Define the sample range $U = \max_i X_i - \min_i X_i$. Then the probability density of U is

$$f_U(u) = n(n-1) \int_a^{b-u} [F(u+z) - F(z)]^{n-2} f(u+z) f(z)\, dz, \quad 0 < u < b - a.$$

§3.43. Median of a Sample

Let $X_1, \ldots, X_n$ be independent identically distributed random variables each with a continuous distribution $F(x)$, and a probability density $f(x)$. Suppose $f(x) \neq 0$ only for $a < x < b$ (a and b may be $-\infty$ and $+\infty$). Let $Y_1 \leqslant \cdots \leqslant Y_n$ denote the order statistics associated with $X_1, \ldots, X_n$. The probability density of the median:

$$M = [Y_{n/2} + Y_{(n+2)/2}]/n, \qquad n = \text{even},$$

$$M = Y_{(n+1)/2}, \qquad n = \text{odd},$$

is given by

$$f_M(y) = \frac{n!\,2}{\left[\left(\frac{n}{2} - 1\right)!\right]^2} \int_y^b [f(2y - z)]^{(n/2)-1}[1 - F(z)]^{(n/2)-1} f(2y - z) f(z)\, dz,$$

$n = \text{even}$,

$$f_M(y) = \frac{n!}{\left[\left(\frac{n-1}{2}\right)!\right]^2} [F(y)]^{(n-1)/2}[1 - F(y)]^{(n-1)/2} f(y), \qquad n = \text{odd},$$

$a < y < b$.

§3.44. Extremes of a Sample

Let $X_1, \ldots, X_n$ be independent identically distributed random variables each with a continuous distribution $F(x)$, and probability density $f(x)$. Suppose $f(x) \neq 0$ only for $a < x < b$ (a and b may be $-\infty$ and $+\infty$). The probability densities of the extremes $\min_i X_i$, $\max_i X_i$ are, respectively

$$f_1(x) = n[1 - F(x)]^{n-1} f(x), \qquad a < x < b,$$

$$f_2(x) = n[F(x)]^{n-1} f(x), \qquad a < x < b.$$

§3.45. Studenized Range

Let $X_1, \ldots, X_n$ be independent standard normal random variables, and define the range $U = \max_i X_i - \min_j X_j$. Let χ^2 be a chi-square variable with v degrees of freedom, and suppose that U and χ^2 are independent. The Studenized range, of v degrees of freedom and parameter n, is defined by $H = U/\sqrt{\chi^2/v}$. The probability density of H is given by

$$f_H(h) = \int_0^\infty f_U(h\sqrt{z}) f_{\chi^2}(vz) v \sqrt{z}\, dz, \qquad 0 < h < \infty,$$

where f_U and f_{χ^2} denote the probability densities of U and χ^2, respectively.

§3.46. Probability Integral Transform

Let X be a random variable with a *continuous* distribution $F(x)$. Then the random variable $F(X)$ is uniformly distributed over (0, 1). This distribution-free character, however, is not true if F depends on some unknown parameters, and the latter are estimated from a sample (see §2.18). We give explicit expressions (David and Johnson) of the corresponding densities for the normal exponential distributions when the parameters are estimated by a sample.

Normal Distribution. Let $F(x; \mu, \sigma^2)$ denote the normal distribution with unknown mean μ and unknown variance σ^2. Let $X_1, \ldots, X_n$ denote independent identically, distributed random variables each with a distribution $F(x; \mu, \sigma^2)$. We estimate μ and σ^2 by the sample mean $\bar{X} = \sum_{i=1}^n X_i/n$ and the sample variance $S^2 = \sum_{i=1}^n (X_i - \bar{X})^2/(n-1)$, respectively. Define the statistic $F(X_j; \bar{X}, S^2)$ by formally replacing μ and σ^2 by $\bar{X}$ and S^2 in $F(\cdot; \mu, \sigma^2)$. The probability density $h(y_i)$ of the statistic $F(X_j; \bar{X}, S^2)$ is given by

$$h(y_j) = \frac{\sqrt{2n}}{(n-1)} \frac{\Gamma\left(\dfrac{n-1}{2}\right)}{\Gamma\left(\dfrac{n-2}{2}\right)} \left[1 - \frac{nz_j^2}{(n-1)^2}\right]^{(n-4)/2} \exp\left[\frac{z_j^2}{2}\right], \qquad j = 1, \ldots, n,$$

where the auxiliary variable z_j is defined through:

$$y_j = (2\pi)^{-1/2} \int_{-\infty}^{z_j} e^{-t^2/2}\, dt, \qquad -(n-1)/\sqrt{n} < z_j < +(n-1)/\sqrt{n}.$$

For $j =$ fixed, $n \to \infty$, $h(y_j)$ approaches the density of the uniform distribution on (0, 1).

Exponential Distribution. Let $X_1, \ldots, X_n$ be independent identically distributed random variables each with an exponential distribution $F(x; \alpha)$ with probability density $f(x) = \alpha e^{-\alpha x}$, $x > 0$. *If* one estimates $1/\alpha$ by $\bar{X}$, then the probability density $h(y_j)$ of the statistic $F(X_j; 1/\bar{X}) = 1 - \exp - (X_j/\bar{X})$ is given by:

$$h(y_j) = \left(\frac{n-1}{n}\right) \frac{1}{(1-y_j)} \left[1 + \frac{\ln(1-y_j)}{n}\right]^{n-2}, \qquad 0 < y_j < 1 - e^{-n},$$

for $j = 1, \ldots, n$. For fixed j, $n \to \infty$, $h(y_j)$ approaches the density of the uniform distribution on (0, 1).

[Cf., David and Johnson (1948), Manoukian (1986).]

§3.47. $\bar{X}, \bar{X} - \bar{Y}$

Let $X_1, \ldots, X_{n_1}$ and $Y_1, \ldots, Y_{n_2}$ be independent random variables. The X's are identically distributed each with mean μ_1 and variance σ_1^2 assumed finite. The Y's are identically distributed each with mean μ_2 and variance σ_2^2 assumed finite. Define

$$\sum_{i=1}^{n_1} (X_i - \bar{X})^2/(n_1 - 1) = S_1^2, \qquad \sum_{i=1}^{n_2} (Y_i - \bar{Y})^2/(n_2 - 1) = S_2^2,$$

$$\bar{X} = \sum_{i=1}^{n_1} X_i/n_1, \qquad \bar{Y} = \sum_{i=1}^{n_2} Y_i/n_2.$$

(A) Normal Populations

(i) $\sqrt{n_1}(\bar{X} - \mu_1)/\sigma_1$ has a standard normal distribution.

(ii) $\sqrt{n_1}(\bar{X} - \mu_1)/S_1$ has a Student distribution of $(n_1 - 1)$ degrees of freedom.

(iii)
$$\frac{[\bar{X} - \bar{Y} - (\mu_1 - \mu_2)]}{\sqrt{\dfrac{\sigma_1^2}{n_1} + \dfrac{\sigma_2^2}{n_2}}}$$

has a standard normal distribution.

(iv) If $\sigma_1^2 = \sigma_2^2$,

$$\frac{[\bar{X} - \bar{Y} - (\mu_1 - \mu_2)]}{\sqrt{\dfrac{1}{n_1} + \dfrac{1}{n_2}}\sqrt{\dfrac{(n_1 - 1)S_1^2 + (n_2 - 1)S_2^2}{n_1 + n_2 - 2}}}$$

has a Student distribution of $n_1 + n_2 - 2$ degrees of freedom.

(v) If $\sigma_1^2 \neq \sigma_2^2$,

$$\frac{[\bar{X} - \bar{Y} - (\mu_1 - \mu_2)]}{\sqrt{\dfrac{S_1^2}{n_1} + \dfrac{S_2^2}{n_2}}}$$

has an approximate Student distribution of

$$v = \frac{\left(\dfrac{s_1^2}{n_1} + \dfrac{s_2^2}{n_2}\right)^2}{\left(\dfrac{s_1^4}{n_1^2(n_1 - 1)} + \dfrac{s_2^4}{n_2^2(n_2 - 1)}\right)}$$

degrees of freedom, where s_1^2 and s_2^2 denote the values obtained for S_1^2 and S_2^2 from a sample. If v is not a natural number, then an extrapolation of the results is necessary from tables of the Student distribution with $v_1 < v < v_1 + 1$, where v_1 is some positive integer.

(B) Unknown Populations

For n_1 sufficiently large, the statistic in (A)(i) has (approximately) a standard normal distribution. Let $\gamma_2^{(1)}$ and $\gamma_2^{(2)}$ denote the kurtosis of the X and Y

distributions, respectively, and suppose $0 < 2 + \gamma_2^{(i)} < \infty$, for $i = 1, 2$. Then for n_1, n_2 sufficiently large, each of the statistics in (ii)–(v) has (approximately) a standard normal distribution.

Remarks. For normal populations, the statistics $\sqrt{n_1}(\bar{X} - \Delta_1)/S_1$, and (for $\sigma_1^2 = \sigma_2^2$)

$$\frac{(\bar{X} - \bar{Y} - \Delta)}{\left(\sqrt{\frac{1}{n_1} + \frac{1}{n_2}}\sqrt{\frac{(n_1 - 1)S_1^2 + (n_2 - 1)S_2^2}{n_1 + n_2 - 2}}\right)},$$

with $\Delta_1 \neq \mu_1$, $\Delta \neq \mu_1 - \mu_2$, have noncentral Student distributions of $(n_1 - 1)$ and $(n_1 + n_2 - 2)$ degrees of freedom and noncentrality parameters $\sqrt{n_1}(\mu_1 - \Delta_1)/\sigma_1$ and

$$\frac{(\mu_1 - \mu_2 - \Delta)}{\sigma\sqrt{\frac{1}{n_1} + \frac{1}{n_2}}} \qquad (\sigma_1 = \sigma_2 \equiv \sigma),$$

respectively.

See also §§1.3 and 2.15.

§3.48. S_1^2, S_1^2/S_2^2, and Bartlett's M

Let $X_1, \ldots, X_{n_1}; Y_1, \ldots, Y_{n_2}; \ldots$ be independent random variables. The X's are identically distributed each with mean μ_1, variance σ_1^2 and kurtosis $\gamma_2^{(1)}$; the Y's are identically distributed each with mean μ_2, variance σ_2^2 and kurtosis $\gamma_2^{(2)}$; .... Suppose $\gamma_2^{(1)} = \gamma_2^{(2)} = \cdots \equiv \gamma_2$ and $0 < 2 + \gamma_2 < \infty$. Set

$$\sum_{i=1}^{n_1} (X_i - \bar{X})^2/(n_1 - 1) = S_1^2, \qquad \sum_{i=1}^{n_1} X_i/n_i = \bar{X}, \ldots.$$

(A) Normal Populations

(i) $(n_1 - 1)S_1^2/\sigma_1^2$ has a (χ^2) chi-square distribution of $(n_1 - 1)$ degrees of freedom.

(ii) $(S_1^2/\sigma_1^2)/(S_2^2/\sigma_2^2)$ has a Fisher's F-distribution of $(n_1 - 1)$ and $(n_2 - 1)$ degrees of freedom.

(iii) Bartlett's M: It is defined by

$$M = -\sum_{j=1}^{k} \nu_j \ln S_j^2 + \nu \ln\left(\sum_{j=1}^{k} \frac{\nu_j}{\nu} S_j^2\right),$$

where $\nu_j = n_j - 1$, $\nu = \sum_{j=1}^{k} \nu_j$. For $\sigma_1^2 = \cdots = \sigma_k^2$,

$$M = -\sum_{j=1}^{k} \nu_j \ln\left[\frac{\chi_j^2}{\nu_j}\left(\frac{\nu}{\chi_1^2 + \cdots + \chi_k^2}\right)\right],$$

and its characteristic function is

$$\Phi(t) = \frac{\prod_{j=1}^{k} [\Gamma(z_j/2)/\Gamma(v_j/2)] \exp[itv_j \ln(v_j/v)]}{[\Gamma(z/2)/\Gamma(v/2)]},$$

where

$$z_j = v_j(1 - 2it), \qquad j = 1, \ldots, k,$$
$$z = v(1 - 2it).$$

$\Gamma(z)$ is the gamma function, and χ_i^2 is a chi-square variable with v_i degrees of freedom.

(a) For $\sigma_1^2 = \cdots = \sigma_k^2$, $v_1 = \cdots = v_k \equiv \underline{v}$, the critical values $M_c(\alpha, \underline{v})$ defined through: $P[M > M_c(\alpha, \underline{v})] = \alpha$ have been tabulated for different values of $\underline{v}$ and α. Let $M_c(\alpha, \underline{v}) = -k\underline{v} \ln A^*$, then the values A^* are tabulated in, e.g., Glaser (1976, p. 489), upon identifying $k \equiv n$ and $\underline{v} \equiv d$, in the latter reference.

(b) For $\sigma_1^2 = \cdots = \sigma_k^2$, and $v_1, \ldots, v_k$ not necessarily equal several approximations have been suggested in the literature for $v_1, \ldots, v_k$ large. Here we mention only two. Let $P[M > M_c(\alpha; v_1, \ldots, v_k)] = \alpha$. (1) For $v_1, \ldots, v_k$ large, approximate $M_c(\alpha; v_1, \ldots, v_k)$ by $C\chi_\alpha^2(k-1)$, where

$$C = 1 + \frac{1}{3(k-1)}\left[\left(\sum_{j=1}^{k} \frac{1}{v_j}\right) - \frac{1}{v}\right],$$

and $\chi_\alpha^2(k-1)$ is the $(1-\alpha)$th quantile of a chi-square distribution of $(k-1)$ degrees of freedom [Bartlett]. (2) For $v_1, \ldots, v_k$ large, approximate $M_c(\alpha; v_1, \ldots, v_k)$ by $(C_1/C_2)M_c(\alpha; \underline{v})$ where $\underline{v}$, here, is defined by $\underline{v} = \min(v_1, \ldots, v_k)$, the $M_c(\alpha; \underline{v})$ are the values given in (a) above, and

$$C_1 = 1 + \frac{1}{3(k-1)}\left[\left(\sum_{j=1}^{k} \frac{1}{v_j}\right) - \frac{1}{v}\right],$$

$$C_2 = 1 + \frac{1}{3\underline{v}}\left(1 + \frac{1}{k}\right) \quad \text{[Manoukian (1983)].}$$

(B) Unknown Populations

(i)
$$\sqrt{\frac{n_1}{2}}\left(\frac{S_1^2}{\sigma_1^2} - 1\right)$$

is for n_1 sufficiently large (approximately) distributed as $\sqrt{1 + (\gamma_2/2)}\, Z$, where Z is a standard normal variable.

(ii)
$$\sqrt{\frac{n_1 n_2}{n_1 + n_2}}\left(\frac{S_1^2}{S_2^2} - \frac{\sigma_1^2}{\sigma_2^2}\right)$$

is for n_1 and n_2 sufficiently large (approximately) distributed as a normal random variable with mean 0 and variance $(\sigma_1^4/\sigma_2^4)(2 + \gamma_2)$. For $\gamma_2^{(1)} \neq \gamma_2^{(2)}$ replace the latter variance by

$$\frac{\sigma^4}{\sigma_2^4}\left[\frac{n_2}{n_1 + n_2}(2 + \gamma_2^{(1)}) + \frac{n_1}{n_1 + n_2}(2 + \gamma_2^{(2)})\right].$$

(iii) Bartlett's M. For $\sigma_1^2 = \cdots = \sigma_k^2$, and $v_1, \ldots, v_k$ sufficiently large, M is (approximately) distributed as $(1 + (\gamma_2/2))\chi_{k-1}^2$, where χ_{k-1}^2 has a chi-square distribution of $(k - 1)$ degrees of freedom.

See also §§1.3 and 2.15.

[Cf., Manoukian (1986, 1982, 1983), Cyr and Manoukian (1982), Giguère *et al.* (1982).]

§3.49. Bivariate Normal

The probability density function of the bivariate normal random vector $(X, Y)'$ is defined by

$$f(x, y) = \frac{1}{2\pi\sigma_1\sigma_2\sqrt{1 - \rho^2}} \exp\left[\frac{-Q}{2}\right],$$

where

$$Q = \frac{1}{(1 - \rho^2)}\left[\left(\frac{x - \mu_1}{\sigma_1}\right)^2 - 2\rho\left(\frac{x - \mu_1}{\sigma_1}\right)\left(\frac{y - \mu_2}{\sigma_2}\right) + \left(\frac{y - \mu_2}{\sigma_2}\right)^2\right],$$

$$\mu_1 = E[X], \qquad \sigma_1^2 = \sigma^2(X),$$

$$\mu_2 = E[Y], \qquad \sigma_2^2 = \sigma^2(Y),$$

$$\rho = \frac{E[XY] - \mu_1\mu_2}{\sigma_1\sigma_2} \quad \text{(correlation coefficient).}$$

Set $x - \mu_1 = r\cos\theta$, $y - \mu_2 = r\sin\theta$. For $\theta = \theta_1$ and $\theta = \theta_2$, let R denote the region bounded by the two ellipses $k_1 = Q(r, \theta_1)/2$ and $k_2 = Q(r, \theta_2)/2$. Then (Lowerre)

(A) For $0 < \theta_1 < \theta_2 < \pi$ or $\pi < \theta_1 < \theta_2 < 2\pi$:

$$P[(X, Y)' \in R] = \frac{e^{-k_1} - e^{-k_2}}{2\pi}\left\{\tan^{-1}\left[\frac{\sigma_2^2}{\sqrt{1 - \rho^2}}\left(\frac{\cot\theta_1}{\sigma_1} - \frac{\rho}{\sigma_1}\right)\right]\right.$$

$$\left. - \tan^{-1}\left[\frac{\sigma_2}{\sqrt{1 - \rho^2}}\left(\frac{\cot\theta_2}{\sigma_1} - \frac{\rho}{\sigma_2}\right)\right]\right\}.$$

(B) For $0 < \theta_1 < \pi < \theta_2 < 2\pi$:

$$P[(X, Y)' \in R] = \frac{e^{-k_1} - e^{-k_2}}{2} + \text{(expression in (A)).}$$

[Cf., Lowerre (1983).]

§3.50. Sample Correlation Coefficient

Let $(X_1, Y_1)', \ldots, (X_n, Y_n)'$ be independent identically distributed random vectors. Let $E[X_1] = \mu_1$, $E[Y_1] = \mu_2$, $\sigma^2(X_1) = \sigma_1^2$, $\sigma^2(Y_1) = \sigma_2^2$, $\rho = E[(X_1 - \mu_1)(Y_1 - \mu_2)]/\sigma_1\sigma_2$. Define the sample correlation coefficient:

$$R = \frac{\sum_{i=1}^{n} (X_i - \bar{X})(Y_i - \bar{Y})}{\sqrt{\sum_{i=1}^{n} (X_i - \bar{X})^2 \sum_{j=1}^{n} (Y_i - \bar{Y})^2}},$$

where $\bar{X} = \sum_{i=1}^{n} X_i/n$, $\bar{Y} = \sum_{i=1}^{n} Y_i/n$.

(A) Normal Bivariate Distribution

The probability density of R is given by

$$g(r) \equiv \frac{(n-2)\Gamma(n)}{(n-1)\sqrt{2\pi}\,\Gamma(n-\frac{1}{2})}(1-\rho^2)^{(n-1)/2}(1-r^2)^{(n-4)/2}(1-\rho r)^{(3/2)-n}$$
$$\times F(\tfrac{1}{2}, \tfrac{1}{2}, n - \tfrac{1}{2}; \tfrac{1}{2}(1+\rho r)), \qquad -1 < r < 1,$$

where $F(a, b, c; z)$ is the hypergeometric function:

$$F(a, b, c; z) = \frac{\Gamma(c)}{\Gamma(a)\Gamma(b)} \sum_{k=0}^{\infty} \frac{\Gamma(a+k)\Gamma(b+k)}{\Gamma(c+k)} \frac{z^k}{k!}.$$

For n sufficiently large:

$$E[R] \simeq \rho\left[1 - \frac{(1-\rho^2)}{2n}\right],$$

$$\sigma^2(R) \simeq \frac{(1-\rho^2)^2}{(n-1)},$$

$$\gamma_1 \simeq -6\rho n^{-1/2},$$

$$\gamma_2 \simeq 6(12\rho^2 - 1)n^{-1}.$$

The variable $Z = \frac{1}{2}\ln[(1+R)/(1-R)]$ is called Fisher's transformation of R. For n sufficiently large, Z has approximately a normal distribution with mean $\frac{1}{2}\ln[(1+\rho)/(1-\rho)]$ and variance $(n-3)^{-1}$. For $\rho = 0$, the exact probability densities of R and Z are, respectively:

$$g(r) = \frac{\Gamma\left(\frac{n-1}{2}\right)}{\sqrt{\pi}\,\Gamma\left(\frac{n-2}{2}\right)}(1-r^2)^{(n/2)-2}, \qquad -1 < r < 1,$$

$$h(z) = \frac{2\Gamma(n-2)}{\left[\Gamma\left(\frac{n-2}{2}\right)\right]^2} e^{(n-2)z}(e^{2z}+1)^{2-n}, \qquad -\infty < z < \infty.$$

The latter is also the probability density of $\frac{1}{2}\ln F$, where F is a Fisher's F random variable of $v_1 = n-2$ and $v_2 = n-2$ degrees of freedom. Also for $\rho = 0$, $W = (\sqrt{n-2}\,R)/(\sqrt{1-R^2})$ has a Student distribution of $n-2$ degrees of freedom.

(B) Unknown Bivariate Distribution

Let $Z = (X - \mu_1)/\sigma_1$, $W = (Y - \mu_2)/\sigma_2$. For n sufficiently large, $\sqrt{n}(R - \rho)$ has approximately a normal distribution with mean 0 and variance $v = E[Z^2W^2] - \rho E[Z^3W] - \rho E[ZW^3] - (\rho^2/4)E[(Z^2 + W^2)^2]$, where it is assumed that the involved moments exists.

See also §§1.3 and 2.15.

[Cf., Kendall and Stuart (1977), Serfling (1980), Manoukian (1986).]

§3.51. Multivariate Normal

If the characteristic function associated with a distribution is of the form:

$$\Phi(\mathbf{t}) = \exp\left\{i\left[\mathbf{t}'\boldsymbol{\mu} - \frac{\mathbf{t}'\boldsymbol{\Sigma}\mathbf{t}}{2}\right]\right\},$$

where $\mathbf{t} = (t_1, \ldots, t_k)' \in R^k$,

$$\mathbf{t}'\boldsymbol{\mu} = \sum_{i=1}^{k} \mu_i t_i, \qquad \mathbf{t}'\boldsymbol{\Sigma}\mathbf{t} = \sum_{i=1}^{k}\sum_{j=1}^{k} t_i \Sigma_{ij} t_j,$$

and the quadratic form $\mathbf{t}'\boldsymbol{\Sigma}\mathbf{t}$ is nonnegative definite, the underlying distribution is called a multivariate normal distribution. If $\mathbf{t}'\boldsymbol{\Sigma}\mathbf{t} > 0$ for all $\mathbf{t} \neq 0$, that is $\mathbf{t}'\boldsymbol{\Sigma}\mathbf{t}$ is positive definite, then the underlying distribution is referred to as nonsingular normal. If $\mathbf{t}'\boldsymbol{\Sigma}\mathbf{t} \geqslant 0$ for $\mathbf{t} \neq 0$, that is, $\mathbf{t}'\boldsymbol{\Sigma}\mathbf{t}$ is semi-definite, then it is referred to as singular. When $\mathbf{t}'\boldsymbol{\Sigma}\mathbf{t}$ is positive definite we can invert the matrix $\boldsymbol{\Sigma}$ and the probability density of the multivariate normal (nonsingular) distribution is:

$$f(\mathbf{x}) = (2\pi)^{-k/2}(\det \boldsymbol{\Sigma})^{-1/2} \exp[-\tfrac{1}{2}(\mathbf{x} - \boldsymbol{\mu})'\boldsymbol{\Sigma}^{-1}(\mathbf{x} - \boldsymbol{\mu})], \qquad (*)$$

with $-\infty < x_i < \infty$, $i = 1, \ldots, k$, $\mathbf{X} = (X_1, \ldots, X_k)'$, and det denotes the determinate of a matrix.

$$E[\mathbf{X}] = \boldsymbol{\mu}, \qquad \operatorname{Cov}[X_i, X_j] = \Sigma_{ij}.$$

We denote the distribution associated with the density in (*) by $N(\boldsymbol{\mu}, \boldsymbol{\Sigma})$, where $\boldsymbol{\mu}$ is the mean vector and $\boldsymbol{\Sigma} = [\Sigma_{ij}]$ is the covariance matrix. If $\Sigma_{ij} \propto \delta_{ij}$, where δ_{ij} is the Kronecker delta, that is, $\delta_{ii} = 1$, for $i = 1, \ldots, k$, $\delta_{ij} = 0$ for $i \neq j$, then $X_1, \ldots, X_k$ are independent. [If $\mathbf{X}$ has a $N(\boldsymbol{\mu}, \boldsymbol{\Sigma})$ distribution, then $\mathbf{c}'\mathbf{X}$ has a $N(\mathbf{c}'\boldsymbol{\mu}, \mathbf{c}'\boldsymbol{\Sigma}\mathbf{c})$ distribution, where $\mathbf{c}$ is a k-vector.]

[Cf., Giri (1977), Muirhead (1982), Wilks (1962), Manoukian (1986).]

§3.52. Wishart

Let $\mathbf{X}_1, \ldots, \mathbf{X}_n$, $n > k$, denote independent identically distributed random k-vectors each with a $N(\boldsymbol{\mu}, \boldsymbol{\Sigma})$ distribution. Define

$$\mathbf{V} = \sum_{i=1}^{n} (\mathbf{X}_i - \bar{\mathbf{X}})(\mathbf{X}_i - \bar{\mathbf{X}})',$$

where $\bar{\mathbf{X}} = \sum_{i=1}^{n} \mathbf{X}_i/n$. Then the distribution of $\mathbf{V}$ is called the Wishart distribution of parameter matrix $\boldsymbol{\Sigma}$ and with $(n-1)$ degrees of freedom. For $k = 1$, $\mathbf{V}/(n-1)$ has a chi-square distribution of $(n-1)$ degrees of freedom. *For* $\mathbf{V}$ positive definite the Wishart probability density is given by:

$$\frac{[\det \mathbf{V}]^{(n-k-2)/2} \exp[-\frac{1}{2} \operatorname{Tr} \boldsymbol{\Sigma}^{-1}\mathbf{V}]}{2^{(n-1)k/2}\pi^{k(k-1)/4}(\det \boldsymbol{\Sigma})^{(n-1)/2} \prod_{j=1}^{k} \Gamma(\frac{1}{2}(n-j))},$$

and is 0 otherwise, and where Tr, det denote, respectively, trace and the determinant of a matrix. $\Gamma(z)$ denotes the gamma function. [If $n \leqslant k$ the distribution is singular.]

Let $\mathbf{B}$ be a real $(k \times k)$ matrix, then

$$E[\exp(i \operatorname{Tr} \mathbf{BV})] = (\det \boldsymbol{\Sigma}^{-1})^{(n-1)/2}/(\det(\boldsymbol{\Sigma}^{-1} - 2i\mathbf{B}))^{(n-1)/2}.$$

[Cf., Giri (1977), Muirhead (1982), Wilks (1962).]

§3.53. Hotelling's T^2

Let $\mathbf{X}_1, \ldots, \mathbf{X}_n$, $n > k$, be independent identically distributed random k-vectors, each with a $N(\boldsymbol{\mu}, \boldsymbol{\Sigma})$ distribution. Suppose $\boldsymbol{\mu} = \boldsymbol{\mu}_0$. The Hotelling T^2-statistic is defined by

$$T^2 = n(\bar{\mathbf{X}} - \boldsymbol{\mu}_0)'\mathbf{S}^{-1}(\bar{\mathbf{X}} - \mu_0),$$

where

$$\mathbf{S} = \sum_{i=1}^{n} (\mathbf{X}_i - \bar{\mathbf{X}})(\mathbf{X}_i - \bar{\mathbf{X}})'/(n-1), \qquad \bar{\mathbf{X}} = \sum_{i=1}^{n} \mathbf{X}_i/n.$$

Then $[(n-k)/k(n-1)]T^2$ has a Fisher F-distribution of $v_1 = k$, $v_2 = n-k$ degrees of freedom. If $\boldsymbol{\mu} \neq \boldsymbol{\mu}_0$, then $[(n-k)/k(n-1)]T^2$, where $T^2 = n(\bar{\mathbf{X}} - \boldsymbol{\mu}_0)'\mathbf{S}^{-1}(\bar{\mathbf{X}} - \boldsymbol{\mu}_0)$ defined in terms of $\boldsymbol{\mu}_0$, has a noncentral F-distribution with $v_1 = k$, $v_2 = n-k$ degrees of freedom and a noncentrality parameter

$\delta = n(\boldsymbol{\mu} - \boldsymbol{\mu}_0)'\boldsymbol{\Sigma}^{-1}(\boldsymbol{\mu} - \boldsymbol{\mu}_0)$. [The statistic T^2 is sometimes referred to as having $(n-1)$ degrees of freedom.]

[Cf., Giri (1977), Muirhead (1982).]

§3.54. Dirichlet

The Dirichlet distribution is a k-dimensional generalization of the beta distribution (§3.22). The probability density of a random vector $\mathbf{X} = (X_1, \ldots, X_k)'$, having a Dirichlet distribution, is defined by

$$f(x_1, \ldots, x_k) = \frac{\Gamma(\alpha_1 + \cdots + \alpha_{k+1})}{\Gamma(\alpha_1)\ldots\Gamma(\alpha_{k+1})}(x_1)^{\alpha_1 - 1}\ldots(x_k)^{\alpha_k - 1}(1 - x_1 - \cdots - x_k)^{\alpha_{k+1}-1},$$

where $0 \leqslant x_i \leqslant 1$, $i = 1, \ldots, k$, such that $\sum_{i=1}^{k} x_i \leqslant 1$, and is zero otherwise; $\alpha_i > 0$, $i = 1, \ldots, k+1$. $\Gamma(\alpha)$ is the gamma function. For $k = 1$, we have the beta distribution with $\alpha_1 = \alpha$, $\alpha_2 = \beta$.

$$E[X_i] = \alpha_i(\alpha_1 + \cdots + \alpha_{k+1})^{-1},$$

$$\sigma^2(X_i) = \alpha_i(\alpha_1 + \cdots + \alpha_{k+1} - \alpha_i)(\alpha_1 + \cdots + \alpha_{k+1})^{-2}(\alpha_1 + \cdots + \alpha_{k+1} + 1)^{-1},$$

for $i = 1, \ldots, k$,

$$\text{Cov}[X_i, X_j] = -\alpha_i\alpha_j(\alpha_1 + \cdots + \alpha_{k+1})^{-2}(\alpha_1 + \cdots + \alpha_{k+1} + 1)^{-1}, \qquad i \neq j,$$

with i, j in $(1, \ldots, k)$.

[Cf., Wilks (1962).]

CHAPTER 4

Some Relations Between Distributions

§4.1. Binomial and Binomial

If $X_1, \ldots, X_k$ are independent random variables having binomial distributions with parameters $(n_1, p), \ldots, (n_k, p)$, respectively, then $\sum_{i=1}^{k} X_i$ has a binomial distribution with parameters (n, p), where $n = \sum_{i=i}^{k} n_i$.

§4.2. Binomial and Multinomial

For $k = 2$, the multinomial distribution reduces to the binomial one. More interestingly, if $X_1, \ldots, X_k$ have a joint multinomial distribution of parameters $(n, p_1, \ldots, p_k)$, then

$$P[X_i = x_i] = b(x_i; n, p_i), \qquad i = 1, \ldots, k.$$

§4.3. Binomial and Beta

If $X(\alpha, \beta)$ has a beta distribution with parameters $\alpha = i$ and $\beta = n - i + 1$, then

$$B(n - i; n, 1 - x) = P[X(\alpha, \beta) \leqslant x],$$

where n, i, with $n \geqslant i$, are natural numbers, and $0 < x < 1$.

§4.4. Binomial and Fisher's *F*

If $F(\nu_1, \nu_2)$ has a Fisher F-distribution of ν_1 and ν_2 degrees of freedom, then

$$B(x; n, p) = 1 - P\left[F(\nu_1, \nu_2) < \frac{n-x}{x+1} \cdot \frac{p}{1-p}\right],$$

where $\nu_1 = 2(x+1)$, $\nu_2 = 2(n-x)$, n, x, with $n > x$, are natural numbers, and $0 < p < 1$.

§4.5. Binomial and Hypergeometric

For $N \to \infty$, $D/N \to p$, the hypergeometric distribution may be approximated by the binomial one with parameters n and p. In practice, it is often recommended that we carry out the approximation if $n < 0.1N$.

§4.6. Binomial and Poisson

For $n \to \infty$, $p \to 0$, with $np \to \lambda \neq 0$, the binomial distribution may be approximated by the Poisson one with parameter λ. In practice, it is often recommended that we carry out the approximation if $p < 0.5$ and $n > 20$.

§4.7. Binomial and Normal

For $n \to \infty$, $p \to \neq 0, \neq 1$, the binomial distribution may be approximated by the normal one with mean $\mu = np$ and variance $\sigma^2 = np(1-p)$. In practice, it is often recommended that we carry out the approximation if $np(1-p) > 9$. [Since $p(1-p) \leqslant \frac{1}{4}$, the inequality $np(1-p) > 9$ implies $n > 36$.] If $X(n, p)$ is a random variable having a binomial distribution with parameters n and p, then the approximation consists of:

$$P[a \leqslant X(n, p) \leqslant b] \simeq \phi(b') - \phi(a'),$$

where

$$b' = \frac{b - np}{\sqrt{np(1-p)}}, \qquad a' = \frac{a - np}{\sqrt{np(1-p)}},$$

$$\phi(a') = \frac{1}{\sqrt{2\pi}} \int_{-\infty}^{a'} e^{-x^2/2}\, dx.$$

In general, a better approximation is given by:

$$P[a \leqslant X(n, p) \leqslant b] \simeq \phi(b^*) - \phi(a^*),$$

where

$$b^* = \frac{b + 0.5 - np}{\sqrt{np(1-p)}}, \qquad a^* = \frac{a - 0.5 - np}{\sqrt{np(1-p)}}.$$

§4.8. Geometric and Pascal

If $X_1(p), \ldots, X_s(p)$ are s independent random variables, having each a geometric distribution of parameter p, then $\sum_{i=1}^{s} X_i(p)$ has a Pascal distribution of parameters s and p.

§4.9. Beta and Beta

If $X(\alpha, \beta)$ has a beta distribution of parameters α and β, then we have the following symmetry property:

$$P[X(\alpha, \beta) \leqslant x] = P[X(\beta, \alpha) > 1 - x], \qquad 0 < x < 1.$$

§4.10. Beta and Fisher's F

If $X(\alpha, \beta)$ has a beta distribution with parameters α and β, and $F(\nu_1, \nu_2)$ has a Fisher F-distribution of ν_1 and ν_2 degrees of freedom, then

$$P\left[X(\alpha, \beta) \leqslant \frac{\alpha}{\alpha + \beta x}\right] = P[F(2\beta, 2\alpha) > x],$$

where 2α and 2β are natural numbers.

§4.11. Beta and Chi-Square

If $\chi_1^2(2\alpha)$ and $\chi_2^2(2\beta)$ are two independent random variables having chi-square distributions of 2α and 2β degrees of freedom, respectively, and $X(\alpha, \beta)$ has a beta distribution with parameters α and β, then

$$P[X(\alpha, \beta) \leqslant x] = P\left[\frac{\chi_1^2(2\alpha)}{\chi_1^2(2\alpha) + \chi_2^2(2\beta)} \leqslant x\right].$$

§4.12. Beta and Uniform

If $X(\alpha, \beta)$ has a beta distribution with parameters α and β, and $Y(a, b)$ has a uniform distribution of parameters a and b, then $X(1, 1)$ and $Y(0, 1)$ have the same distribution.

§4.13. Poisson and Poisson

If $X_1(\lambda_1), \ldots, X_k(\lambda_k)$ are independent random variables having Poisson distributions with parameters $\lambda_1, \ldots, \lambda_k$, respectively, then $X(\lambda) = \sum_{i=1}^k X_i(\lambda_i)$ has a Poisson distribution with parameter $\lambda = \sum_{i=1}^k \lambda_i$.

§4.14. Poisson and Chi-Square

If $X(\lambda)$ has a Poisson distribution with parameter λ, and $\chi^2(v)$ has a chi-square distribution with v degrees of freedom, then

$$P[X(\lambda) \leqslant x] = 1 - P[\chi^2(2(x+1)) < 2\lambda],$$

where x is a positive integer.

§4.15. Poisson and Exponential

Let $X(\alpha t)$ denote a random variable having a Poisson distribution with parameter $\lambda = \alpha t$(the mean). If $X(\alpha t)$ represents, for example, the number of customers arriving at a service station in a certain time interval of length t, then the distribution of the interrarrival time between one customer and another is that of the exponential one with variable $T = t$ and mean $= \alpha^{-1}$.

§4.16. Poisson and Normal

For $\lambda \to \infty$, the Poisson distribution may be approximated by a normal one with mean $\mu = \lambda$ and variance $\sigma^2 = \lambda$. In practice it is often recommended that we carry out the approximation if $\lambda \geqslant 20$. If $X(\lambda)$ has a Poisson distribution with parameter λ, then the approximation consists of:

$$P[a \leqslant X(\lambda) \leqslant b] \simeq \phi(b') - \phi(a'),$$

where

$$b' = \frac{b-\lambda}{\sqrt{\lambda}}, \qquad a' = \frac{a-\lambda}{\sqrt{\lambda}}, \qquad \phi(a') = \frac{1}{\sqrt{2\pi}} \int_{-\infty}^{a'} e^{-x^2/2}\, dx.$$

In general, a better approximation is given by:

$$P[a \leqslant X(\lambda) \leqslant b] \simeq \phi(b^*) - \phi(a^*),$$

where

$$b^* = \frac{b + 0.5 - \lambda}{\sqrt{\lambda}}, \qquad a^* = \frac{a - 0.5 - \lambda}{\sqrt{\lambda}}.$$

§4.17. Exponential and Exponential

If $X_1(\alpha_1), \ldots, X_k(\alpha_k)$ are independent random variables having exponential distributions with means $\alpha_1^{-1}, \ldots, \alpha_k^{-1}$, respectively, then $X(\alpha) = \min[X_1(\alpha_1), \ldots, X_k(\alpha_k)]$ has an exponential distribution with mean α^{-1}, where $\alpha = \sum_{i=1}^k \alpha_i$.

§4.18. Exponential and Erlang

If $X_1(\alpha), \ldots, X_k(\alpha)$ are independent random variables each having an exponential distribution with mean α^{-1}, then $X(\alpha, k) = \sum_{i=1}^k X_i(\alpha)$, has an Erlang distribution with parameters α and k. The exponential distribution is a special case of the Erlang one with $k = 1$.

§4.19. Exponential and Weibull

The exponential distribution is a special case of the Weibull one with $\gamma = 0$, $\beta = 1$, and $\alpha \to 1/\alpha$.

§4.20. Exponential and Uniform

If $X(a, b)$ has a uniform distribution with parameters a and b, then $Y = -(b - a)^{-1} \ln[(b - X)/(b - a)]$ has an exponential distribution with parameter $\alpha = b - a$.

§4.21. Cauchy and Normal

If Z_1 and Z_2 are independent standard normal random variables, then Z_1/Z_2 has a Cauchy distribution with parameter $a = 0$ and $b = 1$.

§4.22. Cauchy and Cauchy

If $X_1(a_1, b_1), \ldots, X_k(a_k, b_k)$ are independent random variables having Cauchy distributions with parameters $(a_1, b_1), \ldots, (a_k, b_k)$, respectively, then $X(a, b) = \sum_{i=1}^k X_i(a_i, b_i)$ has a Cauchy distribution with parameters $a = \sum_{i=1}^k a_i$ and $b = \sum_{i=1}^k b_i$.

§4.23. Normal and Lognormal

If X has a normal distribution with mean μ and variance σ^2, then $Y = e^X$ has a lognormal distribution with parameters μ and σ^2.

§4.24. Normal and Normal

If $X_1, \ldots, X_n$ are independent random variables having normal distributions with means $\mu_1, \ldots, \mu_n$ and variances $\sigma_1^2, \ldots, \sigma_n^2$, respectively, then $Y = \sum_{i=1}^n a_i X_i$ has a normal distribution with mean $\mu = \sum_{i=1}^n a_i \mu_i$ and variance $\sigma^2 = \sum_{i=1}^n a_i^2 \sigma_i^2$, where $a_1, \ldots, a_n$ are any real numbers such that at least one of them is not zero.

§4.25. Normal and Chi-Square

(i) If $Z_1, \ldots, Z_v$ are independent standard normal random variables, then $\sum_{i=1}^v Z_i^2$ has a chi-square distribution with v degrees of freedom.

(ii) If $\chi^2(v)$ has a chi-square distribution with v degrees of freedom, then for $v \to \infty$, we may approximate the distribution of $\sqrt{2\chi^2(v)}$ by a normal one with mean $\sqrt{2v - 1}$ and variance 1. Also for $v \to \infty$, the distribution of $(\chi^2(v)/v)^{1/3}$ may be approximated by a normal one with mean $[1 - (2/9v)]$ and variance $(2/9v)$. In practice, it is often recommended that we carry out the approximations only if $v > 30$. For $v \to \infty$, the distribution of $\chi^2(v)$ may be also approximated by a normal one of mean v and variance $2v$. However, the latter approximation is not usually as good as the latter two. The second one provides, in general, a better approximation than the other two.

§4.26. Normal and Multivariate Normal

Let $\mathbf{X}$ denote a random k-vector such that for every k-vector $\mathbf{t}$, $\mathbf{t}'\mathbf{X}$ has a normal distribution with mean $\sum_{i=1}^k t_i \mu_i$ and variance $\sum_{i=1}^k \sum_{j=1}^k t_i \Sigma_{ij} t_j$, then $\mathbf{X}$ has a multivariate normal distribution with mean vector $\boldsymbol{\mu} = (\mu_1, \ldots, \mu_k)'$ and covariance matrix $\boldsymbol{\Sigma} = [\Sigma_{ij}]$.

§4.27. Normal and Other Distributions

If $X_1, \ldots, X_n$ are independent identically distributed random variables each with mean μ and variance σ^2, *where* $-\infty < \mu < \infty$, $0 < \sigma^2 < \infty$, then $\sqrt{n}(\bar{X} - \mu)/\sigma$ has, for $n \to \infty$, a standard normal distribution. [This is the simplest form of the so-called central limit theorem. For other related limit theorems see Chapter 2.] In practice, it is often recommended that we carry out the approximation in question only if $n > 30$ if the distribution of X_1 is symmetric, and only if $n > 60$ if the distribution of X_1 is nonsymmetric.

§4.28. Exponential Family and Other Distributions

Some distributions which belong to the exponential family are:

- *Binomial with Parameter* p: $0 < p < 1$

$$\begin{aligned} b(x; n, p) &= a(p) \exp[b(p)t(x)]h(x), \\ t(x) &= x, \\ a(p) &= (1-p)^n, \\ b(p) &= \ln[p/(1-p)], \\ h(x) &= \binom{n}{x} I_S(x), \end{aligned}$$

where $S = \{0, 1, \ldots, n\}$ and $I_S(x)$ is the indicator function of the set S, that is $I_S(x) = 1$ if $x \in S$, and $I_S(x) = 0$ if $x \notin S$.

- *Poisson with Parameter* λ: $\lambda > 0$

$$\begin{aligned} f(x; \lambda) &= c(\lambda) \exp[b(\lambda)t(x)]h(x), \\ t(x) &= x, \\ c(\lambda) &= e^{-\lambda}, \\ b(\lambda) &= \ln \lambda, \\ h(x) &= (1/x!)I_Z(x), \end{aligned}$$

where $Z = \{0, 1, 2, \ldots\}$.

- *Exponential with Parameter* α: $\alpha > 0$

$$\begin{aligned} f(x; \alpha) &= c(\alpha) \exp[b(\alpha)t(x)]h(x), \\ t(x) &= x, \\ c(\alpha) &= \alpha, \\ b(\alpha) &= -\alpha, \\ h(x) &= I_S(x), \end{aligned}$$

where $S = \{0 < x < \infty\}$.

• *Normal with Mean μ and Variance* 1: $-\infty < \mu < \infty$

$$f(x;\mu) = a(\mu)\exp[b(\mu)t(x)]h(x), \qquad -\infty < x < \infty,$$
$$t(x) = x,$$
$$a(\mu) = \frac{1}{\sqrt{2\pi}}e^{-\mu^2/2},$$
$$b(\mu) = \mu,$$
$$h(x) = e^{-x^2/2}.$$

• *Normal with Mean* 0 *and Variance σ^2*: $0 < \sigma^2 < \infty$

$$f(x;\sigma^2) = c(\sigma^2)\exp[b(\sigma^2)t(x)]h(x), \qquad -\infty < x < \infty,$$
$$t(x) = x^2,$$
$$c(\sigma^2) = 1/\sqrt{2\pi\sigma^2},$$
$$b(\sigma^2) = -1/2\sigma^2,$$
$$h(x) = 1.$$

• *Normal with Mean μ and Variance σ^2*: $\theta_1 = \mu$, $\theta_2 = \sigma^2$, $\boldsymbol{\theta} = (\theta_1\theta_2)'$, $-\infty < \theta_1 < \infty$, $0 < \theta_2 < \infty$,

$$f(x;\boldsymbol{\theta}) = c(\boldsymbol{\theta})\exp\left[\sum_{i=1}^{k} c_i(\boldsymbol{\theta})t_i(x)\right]h(x), \qquad n = 1, \quad k = 2, \quad -\infty < x < \infty,$$
$$c(\boldsymbol{\theta}) = \frac{1}{\sqrt{2\pi\theta_2}}\exp[-\theta_1^2/2\theta_2],$$
$$h(x) = 1,$$
$$t_1(x) = x, \qquad c_1(\boldsymbol{\theta}) = \theta_1/\theta_2,$$
$$t_2(x) = x^2, \qquad c_2(\boldsymbol{\theta}) = -1/2\theta_2.$$

§4.29. Exponential Power and Other Distributions

For $\delta = 1$, the exponential power distribution reduces to the double-exponential one with parameters $\theta = \mu$ and $\beta = \sigma/\sqrt{2}$. For $\delta = \frac{1}{2}$, the exponential power distribution reduces to the normal one with parameters μ and σ^2. For $\delta \to 0$, the exponential power distribution reduces to the uniform one with parameters $b = \mu + \sqrt{3}\,\sigma$, and $a = \mu - \sqrt{3}\,\sigma$.

§4.30. Pearson Types and Other Distributions

Some examples relating the Pearson types to other distributions follow. The beta distribution is of Type I. Also if U has a beta distribution, then the distribution of $U(1 - U)^{-1}$ is of Type VI. The gamma distribution is of Type III. The Pareto distribution is of Type VI, and the Student distribution is of Type VII. For $C_1 = 0$, $C_2 = 0$, $C_0 > 0$, we recover the normal distribution with mean $\mu = 0$ and variance $\sigma^2 = C_0$, and the latter is not assigned to any particular type.

§4.31. Chi-Square and Chi-Square

If $\chi_1^2, \ldots, \chi_k^2$ are independent random variables having chi-square distributions with $\nu_1, \ldots, \nu_k$ degrees of freedom, respectively, then $\chi^2 = \sum_{i=1}^k \chi_i^2$ has a chi-square distribution with $\nu = \sum_{i=1}^k \nu_i$ degrees of freedom.

§4.32. Chi-Square and Gamma

The chi-square distribution is a special case of the gamma distribution with parameters $\alpha = \frac{1}{2}$ and $k = \nu/2$.

§4.33. Chi-Square and Fisher's F

(i) If χ_1^2 and χ_2^2 are independent random variables having chi-square distributions of ν_1 and ν_2 degrees of freedom, then $(\chi_1^2/\nu_1)/(\chi_2^2/\nu_2)$ has a Fisher F-distribution of ν_1 and ν_2 degrees of freedom.

(ii) If χ^2 has a chi-square distribution of ν degrees of freedom, then χ^2/ν has the same distribution as the random variable $F(\nu_1, \nu_2)$ having a Fisher F-distribution with $\nu_1 = \nu$ and $\nu_2 = \infty$.

§4.34. Student, Normal, and Chi-Square

If Z and χ^2 are independent random variables having a standard normal distribution and a chi-square distribution of ν degrees of freedom, respectively, then $(Z/\sqrt{\chi^2/\nu})$ has a Student distribution of ν degrees of freedom.

§4.35. Student and Cauchy

For $\nu = 1$, the Student distribution reduces to the Cauchy one with parameters $a = 0$ and $b = 1$.

§4.36. Student and Hyperbolic-Secant

If T has a Student distribution of one degree of freedom, and X has a hyperbolic-secant distribution, of parameters $\mu = 0$, $\sigma^2 = 1$ then

$$P[X \leqslant c] = P[-e^{\pi c} \leqslant T \leqslant e^{\pi c}].$$

§4.37. Student and Fisher's F

If $T(v)$ has a Student distribution of v degrees of freedom, and $F(v_1, v_2)$ has a Fisher F-distribution of v_1 and v_2 degrees of freedom, then

$$P[T(v) \leqslant x] = \tfrac{1}{2}\{1 + P[F(1, v) \leqslant x^2]\}.$$

§4.38. Student and Normal

For $v \to \infty$, the Student distribution may be approximated by the standard normal one. In practice, it is often recommended that we carry out the approximation if $v \geqslant 30$.

§4.39. Student and Beta

If $T(v)$ has a Student distribution of v degrees of freedom, then $Y(\frac{1}{2}, v/2) = [1 + (v/T^2(v))]^{-1}$ has a beta distribution with parameters $\alpha = \frac{1}{2}$ and $\beta = v/2$.

§4.40. Student and Sample Correlation Coefficient

If R denotes the sample correlation coefficient from a bivariate normal distribution with correlation coefficient $\rho = 0$, then $\sqrt{n - 2R}/\sqrt{1 - R^2}$ has a Student distribution of $(n - 2)$ degrees of freedom, where n is the sample size.

§4.41. Fisher's F and Logistic

If $F(v_1, v_2)$ has a Fisher F-distribution of v_1 and v_2 degrees of freedom, then $\frac{1}{2} \ln F(2, 2)$ has a logistic distribution with parameters $\theta = 0$ and $\beta = 1$.

§4.42. Fisher's F and Fisher's Z-Transform

Let $Z = \frac{1}{2} \ln[(1 + R)/(1 - R)]$, where R is the sample correlation coefficient from a bivariate normal distribution with correlation coefficient $\rho = 0$. Then Z has the same distribution as $\frac{1}{2} \ln F(n - 2, n - 2)$, where $F(v_1, v_2)$ has a

Fisher F-distribution of $\nu_1 = n - 2$ and $\nu_2 = n - 2$ degrees of freedom, and n denotes the sample size.

§4.43. Noncentral Chi-Square and Normal

If $X_1, \ldots, X_\nu$ are independent random variables having normal distributions with means $\mu_1, \ldots, \mu_\nu$ and variances $\sigma_1^2, \ldots, \sigma_\nu^2$, respectively, then $\sum_{i=1}^{\nu} X_i^2/\sigma_i^2$ has a noncentral chi-square distribution of ν degrees of freedom and a noncentrality parameter $\delta = \sum_{i=1}^{\nu} \mu_i^2/\sigma_i^2$.

§4.44. Noncentral Chi-Square and Noncentral Chi-Square

If $\chi_1^2(\delta_1), \ldots, \chi_k^2(\delta_k)$ are independent random variables having noncentral chi-square distributions of $\nu_1, \ldots, \nu_k$ degrees of freedom and noncentrality parameters $\delta_1, \ldots, \delta_k$, then $\chi^2(\delta) = \sum_{i=1}^{k} \chi_i^2(\delta_i)$ has a noncentral chi-square distribution of $\nu = \sum_{i=1}^{k} \nu_i$ degrees of freedom and a noncentrality parameter $\delta = \sum_{i=1}^{k} \delta_i$.

§4.45. Noncentral Student, Normal, and Chi-Square

If X and χ^2 are independent random variables having a normal distribution with mean μ and variance 1, and a chi-square distribution of ν degrees of freedom, respectively, then $X/\sqrt{\chi^2/\nu}$ has a noncentral Student distribution of ν degrees of freedom and a noncentrality parameter μ.

§4.46. Noncentral Fisher's F, Noncentral Chi-Square, and Chi-Square

If $\chi_1^2(\delta)$ and χ_2^2 are independent random variables having a noncentral chi-square distribution of ν_1 degrees of freedom and a noncentrality parameter δ, and a chi-square distribution of ν_2 degrees of freedom, respectively, then $(\chi_1^2(\delta)/\nu_1)/(\chi_2^2/\nu_2)$ has a noncentral Fisher F-distribution of ν_1 and ν_2 degrees of freedom and a noncentrality parameter δ.

§4.47. Multivariate Normal and Multivariate Normal

If $\mathbf{X}_1, \ldots, \mathbf{X}_n$ are independent identically distributed random variables having each a $N(\boldsymbol{\mu}, \boldsymbol{\Sigma})$ distribution, then $\sqrt{n}(\bar{\mathbf{X}} - \boldsymbol{\mu})$ has a $N(0, \boldsymbol{\Sigma})$ distribution, where $\bar{\mathbf{X}} = \sum_{i=1}^{n} \mathbf{X}_i/n$.

§4.48. Multivariate Normal and Chi-Square

(i) If $\mathbf{X}_1, \ldots, \mathbf{X}_n$ are independent random k-vectors having each a $N(\boldsymbol{\mu}, \boldsymbol{\Sigma})$ distribution, then $n(\bar{\mathbf{X}} - \boldsymbol{\mu})'\boldsymbol{\Sigma}^{-1}(\bar{\mathbf{X}} - \boldsymbol{\mu})$ has a chi-square distribution of k degrees of freedom.

(ii) Define the statistic

$$\lambda = (e/n)^{kn/2}(\det \mathbf{V})^{n/2}(\det \boldsymbol{\Sigma})^{-n/2} \exp[-\tfrac{1}{2} \operatorname{Tr}(\boldsymbol{\Sigma}^{-1}\mathbf{V})],$$

where $\mathbf{V} = \sum_{i=1}^n (\mathbf{X}_i - \bar{\mathbf{X}})(\mathbf{X}_i - \bar{\mathbf{X}})'$ where det and Tr denote the determinant and trace of a matrix, respectively. Then for $n \to \infty$, $-2 \ln \lambda$ has a limiting chi-square distribution of $[k(k+1)/2]$ degrees of freedom.

§4.49. Multivariate Normal and Noncentral Chi-Square

(i) Suppose $\mathbf{X} = (X_1, \ldots, X_k)'$ has a multivariate (singular) normal distribution with mean vector $\boldsymbol{\mu} = (\mu_1, \ldots, \mu_k)'$ and covariance matrix $\boldsymbol{\Sigma}$, with $\Sigma_{ij} = \delta_{ij} - \sqrt{\lambda_i \lambda_j}$, where $0 < \lambda_i < 1$, $i = 1, \ldots, k$, $\sum_{i=1}^k \lambda_i = 1$, and

$$\sum_{i=1}^{k} \sqrt{\lambda_i}\, \mu_i = 0.$$

Here δ_{ij} denotes the Kronecker delta: $\delta_{ii} = 1$, $i = 1, \ldots, k$, and $\delta_{ij} = 0$ for $i \neq j$. Then $\sum_{i=1}^k X_i^2$ has a noncentral chi-square distribution of $(k-1)$ degrees of freedom and a noncentrality parameter $\delta = \sum_{i=1}^k \mu_i^2$.

(ii) Let $\mathbf{X}_1, \ldots, \mathbf{X}_n$ be independent identically distributed random k-vectors having each a $N(\boldsymbol{\mu}, \boldsymbol{\Sigma})$ distribution. Then for $\boldsymbol{\mu}_0 \neq \boldsymbol{\mu}$, $n(\bar{\mathbf{X}} - \boldsymbol{\mu}_0)'\boldsymbol{\Sigma}^{-1}(\bar{\mathbf{X}} - \boldsymbol{\mu}_0)$ has a noncentral chi-square distribution of k degrees of freedom and a noncentrality parameter $\delta = n(\boldsymbol{\mu} - \boldsymbol{\mu}_0)'\Sigma^{-1}(\boldsymbol{\mu} - \boldsymbol{\mu}_0)$, where $\bar{\mathbf{X}} = \sum_{i=1}^k \mathbf{X}_i/n$.

§4.50. Multivariant Normal and Fisher's F

Let $\mathbf{X}_1, \ldots, \mathbf{X}_n$. $n > k$, be independent identically distributed random k-vectors having each a $N(\boldsymbol{\mu}, \boldsymbol{\Sigma})$ distribution. Define

$$\mathbf{S} = \sum_{i=1}^{n} (\mathbf{X}_i - \bar{\mathbf{X}})(\mathbf{X}_i - \bar{\mathbf{X}})'/(n-1), \qquad \bar{\mathbf{X}} = \sum_{i=1}^{n} \mathbf{X}_i/n.$$

Then

$$\frac{(n-k)n}{k(n-1)}(\bar{\mathbf{X}} - \boldsymbol{\mu})'\mathbf{S}^{-1}(\bar{\mathbf{X}} - \boldsymbol{\mu})$$

has a Fisher F-distribution of $v_1 = k$ and $v_2 = n - k$ degrees of freedom.

§4.51. Multivariate Normal and Noncentral Fisher's F

Let $\mathbf{X}_1, \ldots, \mathbf{X}_n$, $n > k$, be independent identically distributed random k-vectors having each a $N(\boldsymbol{\mu}, \boldsymbol{\Sigma})$ distribution. Then for $\boldsymbol{\mu}_0 \neq \boldsymbol{\mu}$,

$$\frac{(n-k)n}{k(n-1)}(\bar{\mathbf{X}} - \boldsymbol{\mu}_0)'\mathbf{S}^{-1}(\bar{\mathbf{X}} - \boldsymbol{\mu}_0)$$

has a noncentral Fisher F-distribution of $v_1 = k$ and $v_2 = n - k$ degrees of freedom and a noncentrality parameter $\delta = n(\boldsymbol{\mu} - \boldsymbol{\mu}_0)'\boldsymbol{\Sigma}^{-1}(\boldsymbol{\mu} - \boldsymbol{\mu}_0)$.

§4.52. Dirichlet and Dirichlet

If the random vector $(X_1, \ldots, X_k)'$ has a Dirichlet distribution with parameters $\alpha_1, \ldots, \alpha_{k+1}$, then the marginal distribution of the random vector $(X_1, \ldots, X_j)'$, for any $1 \leqslant j < k$, is that of a Dirichlet distribution with parameters $\alpha_1, \ldots, \alpha_j, (\alpha_{j+1} + \cdots + \alpha_{k+1})$.

§4.53. Dirichlet and Beta

The beta distribution is a special case of the Dirichlet distribution with $k = 1$, $\alpha_1 = \alpha$, $\alpha_2 = \beta$. Also if the random vector $\mathbf{X} = (X_1, \ldots, X_k)'$ has a Dirichlet distribution with parameters $\alpha_1, \ldots, \alpha_{k+1}$, then $(X_1 + \cdots + X_k)$ has a beta distribution with parameters $\alpha = \alpha_1 + \cdots + \alpha_k$, $\beta = \alpha_{k+1}$.

Bibliography

Barra, J.-R. (1981), *Mathematical Basis of Statistics*, Academic Press, New York.

Bartlett, M. S. (1937), Properties of sufficiency and statistical tests, *Proc. Roy. Soc. London* (Ser. A) **160**, 268–282.

Beaumont, G. P. (1980), *Intermediate Mathematical Statistics*, Chapman & Hall, London.

Berger, J. O. (1980), *Statistical Decision Theory*, Springer-Verlag, New York.

Bickel, P. J. (1965), On some robust estimates of location, *Ann. Math. Statist.* **43**, 847–858.

Bickel, P. J. and Doksum, K. A. (1977), *Mathematical Statistics*, Holden-Day, San Francisco.

Billingsley, P. (1979), *Probability and Measure*, Wiley, New York.

Box, G. E. P. (1953), Nonnormality and tests on variances, *Biometrika* **40**, 318–335.

Box, G. E. P. and Tiao, G. C. (1973), *Bayesian Inference in Statistical Analysis*, Addison-Wesley, Reading, MA.

Bradley, J. V. (1968), *Distribution-Free Statistical Tests*, Prentice-Hall, Englewood Cliffs, NJ.

Bradley, J. V. (1969), A survey of sign tests based on the binomial distribution, *J. Qual. Technol.* **1**, No. 2, 89–101.

Burrill, C. W. (1972), *Measure, Integration, and Probability*, McGraw-Hill, New York.

Chernoff, H. and Savage, I. R. (1958), Asymptotic normality and efficiency of certain nonparametric test statistics, *Ann. Math. Statist.* **29**, 972–994.

Cochran, W. G. (1952), The χ^2 test of goodness of fit, *Ann. Math. Statist.* **23**, 314–345.

Cox, D. R. and Hinkley, D. V. (1974), *Theoretical Statistics*, Chapman & Hall, London.

Cramér, H. (1974), *Mathematical Methods of Statistics*, Princeton University Press, Princeton, NJ.

Cyr, J. L. and Manoukian, E. B. (1982), Approximate critical values for Bartlett's test of homogeneity of variances for unequal sample sizes and errors in estimation, *Commun. Statist.* (Ser A) **11**, 1671–1680.

Darling, D. A. (1957), The Kolmogorov–Smirnov, Cramér–von Mises tests, *Ann. Math. Statist.* **28**, 823–838.

David, H. A. (1981), *Order Statistics*, 2nd edn., Wiley, New York.

David, F. N. and Johnson, N. L. (1948), Probability integral transformation when parameters are estimated from the sample, *Biometrika* **35**, 182–190.

Durbin, J. (1973), *Distribution Theory for Tests Based on the Sample Distribution Function*, Regional Conference Series in Applied Mathematics, Vol. 9, SIAM, Philadelphia.

Edgington, E. S. (1980), *Randomization Tests*, Marcel Dekker, New York.

Efron, B. (1979), Bootstrap methods: another look at the jackknife, *Ann. Statist.* **7**, 1–26.

Efron, B. (1982), *The Jackknife, The Bootstrap and Other Resampling Plans*, Regional Conference Series in Applied Mathematics, Vol. *38*, SIAM, Philadelphia.

Efron, B. (1983), A leisury look at the bootstrap, the jackknife, and cross-validation, *Amer. Statist.* **37**, No. 1, 36–48.

Emerson, J. D. and Simon, G. A. (1979), Another look at the sign test when ties are present: the problem of confidence intervals, *Amer. Statist.* **33**, No. 3, 140–142.

Esséen, C. G. (1956), A moment inequality with an application to the central limit theorem, *Skand. Aktuarietidskr.* **39**, 160–170.

Ferguson, T. S. (1967), *Mathematical Statistics*, Academic Press, New York.

Fernholz, L. T. (1983), *von Mises Calculus for Statistical Functions*, Springer-Verlag, New York.

Fisher, L. and McDonald, J. (1978), *Fixed Effects Analysis of Variance*, Academic Press, New York.

Fisher, R. A. (1951), *The Design of Experiments*, 6th edn., Hafner, New York.

Fisz. M. (1980), *Probability Theory and Mathematical Statistics*, 3rd edn., Krieger, Huntington, NY.

Fourgeaud, C. and Fuchs, A. (1967), *Statistique*, Dunod, Paris.

Glaser, R. E. (1976), Exact critical values for Bartlett's test for homogeneity of variances, *J. Amer. Statist. Ass.* **71**, 488–490.

Gnedenko, B. V. (1966), *The Theory of Probability*, Chelsea, New York.

Gradshteyn, I. S. and Ryzhik, I. M. (1965), *Table of Integrals, Series, and Products*, 4th edn., Academic Press, New York.

Gray, H. L. and Schucany, W. R. (1972), *The Generalized Jackknife Statistic*, Marcel Dekker, New York.

Giguère, J. C. M., Manoukian, E. B. and Roy, J. M. (1982), Maximum absolute error for Bartlett's chi-square approximation, *J. Statist. Comput. Simul.* **15**, 109–117.

Giri, N. C. (1977), *Multivariate Statistical Inference*, Academic Press, New York.

Hájek, J. and Šidák, Z. (1967), *Theory of Rank Tests*, Academic Press, New York.

Hampel, F. R. (1968), Contributions to the Theory of Robust Estimation, Ph.D. Thesis, University of California, Berkeley.

Hampel, F. R. (1971), A general qualitative definition of robustness, *Ann. Math. Statist.* **42**, 1887–1896.

Hartigan, J. A. (1983), *Bayes Theory*, Springer-Verlag, New York.

Hoaglin, D. C., Mosteller, F. and Tukey, J. W. (Eds.) (1983), *Understanding Robust and Exploratory Data Analysis*, Wiley, New York.

Hodges, J. L. Jr. and Lehmann, E. L. (1963), Estimates of location based on rank tests, *Ann. Math. Statist.* **34**, 598–611.

Hoeffding, W. (1948), A class of statistics with asymptotically normal distribution, *Ann. Math. Statist.* **19**, 293–325.

Hogg, R. V. (Ed.) (1978), *Studies in Statistics*, Vol. 19, Mathematical Association of America, Washington, DC.

Hogg, R. V. and Craig, A. T. (1978), *Introduction to Mathematical Statistics*, 4th edn., Macmillan, New York.

Horn, P. S. (1983), Measure for peakedness, *Amer. Statist.* **37**, No. 1, 55—56.

Huber, P. J. (1977), *Robust Statistical Procedures*, Regional Conference Series in Applied Mathematics, Vol. 27, SIAM, Philadelphia.

Huber, P. J. (1981), *Robust Statistics*, Wiley, New York.

Jaeckel, L. A. (1971), Robust estimates of location: symmetry and asymmetric contamination, *Ann. Math. Statist.* **42**, 1020–1034.

Johnson, N. L. and Kotz, S. (1969–1972), *Distributions in Statistics* (in 4 volumes), Wiley, New York.

Kendall, M. Sir (1975), *Rank Correlation Methods*, 4th edn., Charles Griffin, London.

Kendall, M. Sir and Stuart, A. (1977), *The Advanced Theory* of Statistics, Vol. 1, 4th edn., Macmillan, New York.

Kendall, M. Sir and Stuart, A. (1979), *The Advanced Theory of Statistics*, Vol. 2, 4th edn., Macmillan, New York.

Kirmani, S. N. U. A. and Isfahani, E. M. (1983), A note on the moment generating function, *Amer. Statist.* **37**, No. 2, 161.

Kruskal, W. H. (1952), A nonparametric test for the several sample problem, *Ann. Math. Statist.* **23**, 525–540.

Layard, M. W. (1973), Robust large-sample tests for homogeneity of variances, *J. Amer. Statist. Assoc.* **68**, 195–198.

Lehmann, E. L. (1951), Consistency and unbiasedness of certain nonparametric tests, *Ann. Math. Statist.* **22**, 165–179.

Lehmann, E. L. (1959), *Testing Statistical Hypotheses*, Wiley, New York.

Lehmann, E. L. (1963), Robust estimation in analysis of variance, *Ann. Math. Statist.* **34**, 957–966.

Lehmann, E. L. (1975), *Nonparametrics*, Holden-Day, Oakland, CA.

Lehmann, E. L. (1983), *Theory of Point Estimation*, Wiley, New York.

Lilliefors, H. (1967), On the Kolmogorov–Smirnov test for normality with mean and variance unknown, *J. Amer. Statist. Ass.* **62**, 399–402.

Lowerre, J. M. (1983), An integral of the bivariate normal distribution and an application, *Amer. Statist.* **37**, No. 3, 235–236.

Manoukian, E. B. (1982), Bounds on the accuracy of Bartlett's chi-square approximation, *SIAM J. Appl. Math.* **42**, 575–587.

Manoukian, E. B. (1983), Departure of Bartlett's distribution for the homogeneity of variances for unequal sample sizes from that of equal sample sizes, *Metrika*, **30**, 179–194.

Manoukian, E. B. (1984a), Asymptotic distribution of the non-null Bartlett Test statistic for the test of homogeneity of scales with unspecified underlying populations and efficiency of tests, *J. Organ. Behav. Statist.* **1**, No. 1, 33–40.

Manoukian, E. B. (1984b), Asymptotic distribution-free property of probability integral transform with unknown location and scale parameters, *Pub. Inst. Statist. Univ.* (Paris) **29**, No. 1, 59–64.

Manoukian, E. B. (1986), *Mathematical Nonparametric Statistics*, Gordon & Breach, New York (in press).

Mason, D. M. and Schuenemeyer, J. H. (1983), A modified Kolmogorov–Smirnov test sensitive to tail alternatives, *Ann. Statist.* **11**, 933–946.

Massey, F. J. (1950), A note on the estimation of a distribution function by confidence limits, *Ann. Math. Statist.* **21**, 116–119.

Massey, F. J. (1951a), The Kolmogorov–Smirnov tests for goodness of fit, *J. Amer. Statist. Ass* **46**, 68–78.

Massey, F. J. (1951b), The distribution of the maximum deviation between two-sample cummulative step functions, *Ann. Math. Statist.* **22**, 125–128.

Miller, R. G. Jr. (1964), A trustworthy jackknife, *Ann. Math. Statist.* **35**, 1594–1605.

Miller, R. G. Jr. (1968), Jackknifing variances, *Ann. Math. Statist.* **39**, 567–582.

Miller, R. G. Jr. (1974), The jackknife—a review, *Biometrika*, **61**, 1–15.
Miller, R. G. Jr. (1981), *Simultaneous Statistical Inference*, 2nd edn. Springer-Verlag, New York.
Milton, J. S. and Tsokos, C. P. (1976), *Probability Theory with the Essential Analysis*, Addison-Wesley, Reading, MA.
Muirhead, R. J. (1982), *Aspects of Multivariate Statistical Theory*, Wiley, New York.
Noether, G. E. (1955), On a theorem by Pitman, *Ann. Math. Statist.* **26**, 64–68.
Noether, G. E. (1963), Note on the Kolmogorov statistic in the discrete case, *Metrika*, **7**, 115–116.
Petrov, V. V. (1975), *Sums of Independent Random Variables*, Springer-Verlag, New York.
Pitman, E. J. G. (1937a), Significance tests which may be applied to samples from any population, *J. Roy. Statist. Soc.* (Ser. B), **4**, 119–130.
Pitman, E. J. G. (1937b), Significance tests which may be applied to samples from any population. II, The correlation coefficient test, *J. Roy. Statist. Soc.* (Ser. B), **4**, 225–232.
Pitman, E. J. G. (1937c), Significance tests which may be applied to samples from any population. III, The analysis of variance test, *Biometrika* **29**, 322–335.
Pratt, J. W. and Gibbons, J. D. (1981), *Concepts of Nonparametric Theory*, Springer-Verlag, New York.
Puri, M. L. and Sen, P. K. (1971), *Nonparametric Methods in Multivariate Analysis*, Wiley, New York.
Putter, J. (1955), The treatment of ties in some nonparametric tests, *Ann. Math. Statist.* **26**, 368–386.
Quenouille, M. H. (1956), Notes on bias in estimation, *Biometrika* **43**, 353–360.
Ramberg, J. S. and Schmeiser, B. W. (1972), An approximate method for generating symmetric random variables, *Commun. A. C. M.* **15**, 987–990.
Ramberg, J. S. and Schmeiser, B. W. (1974), An approximate method for generating asymmetric random variables, *Commun. A. C. M.* **17**, 78–82.
Randles, R. H. and Wolfe, D. A. (1979), *Introduction to the Theory of Nonparametric Statistics*, Wiley, New York.
Rao, C. R. (1973), *Linear Statistical Inference and Its Applications*, 2nd edn., Wiley, New York.
Rényi, A. (1970), *Foundations of Probability*, Holden-Day, San Francisco.
Rey, J. J. W. (1978), *Robust Statistical Methods*, Springer-Verlag, New York.
Rey, J. J. W. (1983), *Introduction to Robust and Quasi-robust Statistical Methods*, Springer-Verlag, New York.
Roussas, G. G. (1973), *A First Course in Mathematical Statistics*, Addison-Wesley, Reading, MA.
Scheffé, H. (1959), *The Analysis of Variance*, Wiley, New York.
Schmetterer, L. (1974), *Introduction to Mathematical Statistics*, Springer-Verlag, Berlin.
Seber, G. A. F. (1977), *Linear Regression Analysis*, Wiley, New York.
Seber, G. A. F. (1980), *The Linear Hypothesis: A General Theory*, 2nd edn. Macmillan, New York.
Serfling, R. J. (1980), *Approximation Theorems of Mathematical Statistics*, Wiley, New York.
Silvey, S. D. (1975), *Statistical Inference*, Chapman & Hall, London.
Slaketer, M. J. (1965), A comparison of the Pearson chi-square and Kolmogorov goodness-of-fit tests with respect to validity, *J. Amer. Statist. Ass.* **60**, 854–858.
Slaketer, M. J. (1966), Comparative validity of the chi-square and two modified chi-square goodness-of-fit tests for small but equal expected frequencies, *Biometrika* **53**, 619–623.
Tucker, H. G. (1967), *A Graduate Course in Probability*, Academic Press, New York.

Tukey, J. W. (1960a), A survey of sampling from contaminated distributions: in *Contributions to Probability and Statistics*, I. Olkin, Ed., Stanford University Press, Stanford, CA.
Tukey, J. W. (1960b), The Practical Relationship Between the Common Transformations of Percentages or Counts and of Amounts, Technical Report **36**, Statistical Research Group, Princeton University, Princeton, NJ.
van Beeck, P. (1972), An application of Fourier methods to the problem of sharpening the Berry–Esséen inequality, *Z. Wahrsch. verw. Gebiete* **23**, 187–196.
Wald, A. (1947), *Sequential Analysis*, Wiley, New York.
Wetherill, G. B. (1980), *Sequential Methods in Statistics*, Chapman & Hall, London.
Wetherill, G. B. (1981), *Intermediate Statistical Methods*, Chapman & Hall, London.
Wheeler, D. J. (1973), Note on the Kolmogorov–Smirnov statistic for a general discontinuous variable, *Metrika* **20**, 101–102.
Wilks, S. (1962), *Mathematical Statistics*, Wiley, New York.
Yarnold, J. K. (1970), The minimum expectation in χ^2 goodness of fit test and the accuracy of approximation for the null distribution, *J. Amer. Statist. Ass.* **65**, 864–886.
Zacks, S. (1971), *The Theory of Statistical Inference*, Wiley, New York.

Author Index

Subject Index